About the Author

Dr. Dennis William Sciama (November 18, 1926–December 18, 1999) was a major post WWII British physicist who, together with his students, made huge contributions to the field of physics during the twentieth century. Dr. Sciama was a student of the astronomer Fred Hoyle, and sat under the supervision of Paul Dirac at Cambridge University while working towards his PhD, which he received in 1953.

Dr. Sciama taught at Cornell University, King's College London, Harvard, and the University of Texas at Austin, but spent most of his career at Cambridge during the '50s and '60s, and at Oxford during the '70s and early '80s. Dr. Sciama had a broad knowledge base in physics and had a productive career making connections between astronomy and astrophysics. Dr. Sciama wrote about radio astronomy, X-ray astronomy, quasars, interstellar and intergalactic medium, astro particle physics and the nature of dark matter. He is best known for his work on general relativity and black holes.

THE UNITY OF THE UNIVERSE

D. W. SCIAMA

WITH DIAGRAMS BY
JOAN WEDGE

DOVER PUBLICATIONS, INC.
MINEOLA, NEW YORK

Copyright

Copyright © 1959 by D. W. Sciama
All rights reserved.

Bibliographical Note

This Dover edition, first published in 2009, is an unabridged republication of the work originally published by Anchor Books, Doubleday & Company, Inc., Garden City, New York, 1959.

Library of Congress Cataloging-in-Publication Data

Sciama, D. W. (Dennis William), 1926–
    The unity of the universe / D.W. Sciama ; with diagrams by Joan Wedge.
      p. cm.
    Reprint. Originally published: Garden City, N.Y. : Doubleday, 1959.
    ISBN-13: 978-0-486-47205-8
    ISBN-10: 0-486-47205-1
      1. Cosmology. I. Title.
QB981.S392 2009
523.1—dc22

2009016248

Manufactured in the United States by Courier Corporation
47205101
www.doverpublications.com

TO
HERMANN BONDI, THOMAS GOLD, AND
FRED HOYLE

PREFACE

Is the universe a vast collection of more or less independent objects or is it a single unit? This book sets out to show, in language intelligible to the layman, that the universe is indeed a single unit; that the behavior of nearby matter is strongly influenced by distant regions of the universe. It is important to recognize this influence and the physical processes which underlie it, since only then can we hope to understand our own corner of the universe. Moreover, we can reverse the argument, for the behavior of nearby matter contains information about the properties of the universe as a whole, information which can be extracted by scientific considerations. With the recognition of this fact, cosmology has at last reached the position foreshadowed many years ago by William Blake:

> *To see a world in a grain of sand,*
> *And a heaven in a wild flower,*
> *Hold infinity in the palm of your hand,*
> *And eternity in an hour.*

I should perhaps emphasize at the outset that cosmology is a highly controversial subject which contains little or no agreed body of doctrine. Although most of the ideas developed in this book have been borrowed from the writings and conversations of others, the synthesis presented here is my own. This caution does not, however, apply to the first part

PREFACE

of the book, which describes how our present observational picture of the universe has been built up, from its beginnings in early Greek measurements on the size of the solar system to the latest observations of the expanding universe.

The rest of the book is mainly concerned with explaining how the influence of distant matter arises and what its consequences are. No knowledge of physics or astronomy is assumed on the part of the reader,[1] only a willingness to follow an argument to its logical conclusion.

A subsidiary theme of the book concerns the question: Are some features of the universe accidental or can they all be accounted for in theoretical terms? For my part, I believe that the main aim of science should be to diminish the realm of the accidental as much as possible. This question has a bearing on the current controversy about the remote past and future of the expanding universe: has it evolved from a highly condensed initial state, its density constantly decreasing, or does it always look roughly the same, the continual creation of new matter compensating for the expansion? We shall see that if the universe does *not* evolve we can account for many of its features which would otherwise be accidental. Of course the last word on the history of the universe lies with observation, but this word has not yet been spoken.

It is a great pleasure to thank all those who have helped me in one way or another to write this book: Professor William McCrea, who has encouraged me over the years; Peter Long, Adam Parry, Oliver Penrose, and Gordon Squires, who criticized the early drafts; Roger Penrose, who, in addition to commenting on the text, made a large number of valuable suggestions about the illustrations; and, in particular, Joan Wedge, both for her criticisms and for designing and drawing the illustrations.

Above all I owe a great debt of gratitude to Professors Hermann Bondi, Thomas Gold, and Fred Hoyle, who initiated

[1] Except in an occasional footnote and perhaps in Chapter IX, where the argument, though non-technical, is rather elaborate.

PREFACE

me into cosmology and whose ideas have been my main guide in the exploration of this strange realm of thought. To them I dedicate this book.

D. W. SCIAMA

Trinity College,
Cambridge
September 1958.

CONTENTS

Preface   vii
List of Figures   xiii
List of Plates   xvii
Acknowledgments   xix

PART I
THE UNIVERSE IN OBSERVATION

	Introduction	3
I.	The Size of the Solar System	5
II.	The Distance to the Stars	15
III.	The Milky Way	29
IV.	External Galaxies	47
V.	The Expanding Universe	55

PART II
THE UNIVERSE IN THEORY

	Introduction	69
VI.	Olbers' Paradox	71
VII.	Mach's Principle	83
VIII.	The Principle of Equivalence	107
IX.	The Origin of Inertia	115
X.	The Clock Paradox	131

CONTENTS

XI.	The General Theory of Relativity	139
XII.	The History of the Universe	151
XIII.	The Uniqueness of the Universe	161
XIV.	The Formation of Galaxies	169
XV.	The Formation of the Elements	181

Epilogue: The Unity and the Uniqueness of the Universe 201

Index 207

LIST OF FIGURES

1. Eratosthenes' method of measuring the earth's circumference — 6
2. Aristarchus' method of measuring the distance of the moon — 7
3. Aristarchus' method of measuring the distance of the sun — 9
4. Calculation of the sun's diameter — 11
5. Parallax — 12
6. The smallness of stellar parallaxes — 16
7. The inverse square law for the apparent brightness of a star — 16
8. The parallax of a nearby star relative to distant stars — 18
9. The radial and transverse motions of the stars — 20
10. The proper motions of near and distant stars — 22
11. The variation in brightness of the star Delta Cephei — 25
12. The period-luminosity relation for Cepheid variables — 26
13. Thomas Wright's model of the Milky Way — 30
14. Sir William Herschel's model of the Milky Way — 34
15. Shapley's model of the Milky Way — 35
16. Rigid and differential rotation — 37
17. The differential rotation of the Milky Way — 39

LIST OF FIGURES

18. The angular diameters of open clusters of stars — 40
19. The belt of fog in the Milky Way — 42
20. The electromagnetic spectrum — 44
21. The spiral arms of the Milky Way — 45
22. The period-luminosity relation for the two types of Cepheid variables — 52
23. The Doppler effect — 57
24. The radial velocities of galaxies relative to the sun and to the center of the Milky Way — 60
25. Hubble's first velocity-distance relation for galaxies — 61
26. The expansion of the universe as seen from different galaxies — 63
27. Plane and solid angles — 73
28. The relation between the apparent brightness of a star and the angle it subtends at the eye — 73
29. The total brightness of a set of stars — 74
30. The total brightness of a set of stars which fill the field of vision — 75
31. The average length of a line of sight from the earth to the stars — 77
32. Charlier's model of the universe — 79
33. The rate of arrival of photons from a moving star — 81
34. The sun's gravitational force on the earth and Newton's second law of motion — 86
35. A satellite rotating around the earth once every twenty-four hours — 87
36. Centrifugal forces — 89
37. The orbits of balls ejected from the earth — 90
38. Coriolis force — 91
39. The inertial forces acting when the frame of reference accelerates in a straight line — 92
40. A Foucault pendulum — 96

LIST OF FIGURES

41. Inertia and gravitation as experienced by a man in a closed box — 110
42. Einstein's suggested relation between inertial forces and the gravitational forces of the stars — 112
43. The forces exerted by an electric charge — 118
44. The relation between inertial forces and the (charge-like) gravitational forces of the stars — 119
45. The similarity between magnetic and Coriolis forces — 126
46. The Doppler effect in the light emitted by a rocket ship — 134
47. Measuring a heated disc — 140
48. Measuring distances in the presence of gravitation — 143
49. A straight line and a triangle on the surface of a sphere — 145
50. Linear and non-linear addition of forces — 149
51. A parabola — 162
52. The birth of a galaxy — 175
53. The effect of galaxies of different mass on the intergalactic gas — 176
54. The Rutherford-Bohr model of the atom — 183
55. The electric and nuclear forces between two protons — 187
56. The relative abundances of the elements — 194

LIST OF PLATES

- I. The sun's corona
- II. The sun's spectrum
- III. The Small Magellanic Cloud
- IV. The Large Magellanic Cloud
- V. A mosaic photography of the summer Milky Way—northern portion
- VI. A globular cluster of stars
- VII. Clouds of stars in the region of Sagittarius
- VIII. An open cluster of stars
- IX. The Andromeda nebula
- X. Four spiral nebulae
- XI. The Hubble classification of galaxies
- XII. The relation between red-shift and distance for galaxies
- XIII. A pair of colliding galaxies
- XIV. A Foucault pendulum

ACKNOWLEDGMENTS

The author would like to thank the following for permission to reproduce copyright material:

The Mount Wilson and Palomar Observatories. The Harvard College Observatory for Plate III. The Royal Greenwich Observatory for Plate I. The Franklin Institute of Philadelphia for Plate XIV. Professor J. H. Oort for Figure 21. The Cambridge University Press for extracts from the late Sir Arthur Eddington's books:

Space, Time and Gravitation, The Mathematical Theory of Relativity, The Internal Constitution of the Stars, and The Macmillan Co. for extracts from *Stellar Movements and the Structure of the Universe.*

All the photographs reproduced in this book were taken by the Mount Wilson and Palomar observatories, except where otherwise stated in the captions.

PART I

THE UNIVERSE IN OBSERVATION

INTRODUCTION

The naked eye can distinguish the earth, the sun, the moon, the comets, five planets, and six thousand stars. This was man's picture of the universe until 1609, when Galileo first pointed a telescope at the sky. With this instrument he extended our knowledge by observing many stars which are too faint to be seen directly. Thereafter, each increase in size of telescope has revealed the existence of still fainter stars. This has prompted many people to ask the question: Will this process ever stop? Have we any chance of discovering, if not every star, at least the main outlines of the whole universe? Most astronomers believe that with the best modern telescopes they have indeed done just this.

The first reaction to this claim will perhaps be one of excitement that such a fundamental result should have been achieved in our own lifetime. Nevertheless, we must remember that often in the past fundamental scientific problems have wrongly been considered solved. The great French mathematician Lagrange, for instance, was envious of Newton's good fortune: "for in his time the system of the world still remained to be discovered." And toward the end of the nineteenth century most physicists believed that there was nothing left for them to do, except to calculate known quantities to more and more decimal places.

Are modern astronomers any more likely to be right than were their eminent predecessors? Despite the danger of saying so, they believe the answer to be "Yes." There is a special reason for believing that the twentieth-century universe is *the* universe; that further discoveries will add much in detail but will not alter the general picture. This reason will begin to emerge at the end of Part I, but in order to prepare for it we must go back to the early Greek astronomers, who were the first to measure the size of the earth and the distance to the sun and the moon—thus providing man with his first orientation towards the dimensions of the universe.

These early measurements, and their modern counterparts, are described in Chapter I. The subsequent chapters show how the more powerful methods introduced in the last hundred and twenty years have extended our knowledge of distances from the solar system to the stars, and from the stars to distant galaxies. This knowledge has played a key role in the elucidation of what is our goal in these chapters—the structure of the universe as a whole.

CHAPTER I

THE SIZE OF THE SOLAR SYSTEM

> Ah, but a man's reach should exceed his grasp,
> Or what's a heaven for?
> **BROWNING**

Introduction

It is not surprising that it was the Greeks, with their profound understanding of geometrical principles, who were the first to devise methods of measuring the size of the earth and the distance to the sun and the moon. Indeed, their results were not superseded until the eighteenth century, when telescopes had been developed to the point where new methods could be introduced. We shall therefore begin our account of the universe in observation by going back to Greek astronomy.

Greek Times

THE SIZE OF THE EARTH. A rough estimate of the size of the earth was made by Aristotle, but we owe the first accurate value to Eratosthenes (276–194 B.C.). He found that when the sun is overhead at Aswan it is 7 degrees from the vertical in Alexandria (Fig. 1a). This simple observation suffices to determine the circumference of the earth in terms of the distance from Aswan (A) to Alexandria (B) (Fig. 1b). For since the sun is at a great distance from the earth, the angle of 7 degrees measured by Eratosthenes is very nearly the same as the angle AOB subtended by the arc AB at the center of the earth O. Now since 7 degrees is nearly one fiftieth of a complete revolution of 360 degrees, the distance between Aswan and Alexandria must be nearly one fiftieth of the circumference of the earth. Eratosthenes knew that this distance

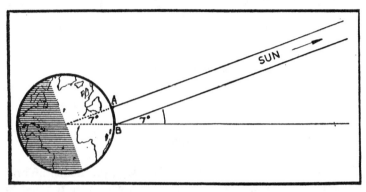

Fig. 1. (a) *Eratosthenes' method of measuring the earth's circumference. When the sun is directly overhead at A, it makes an angle of 7 degrees with the vertical at B.*

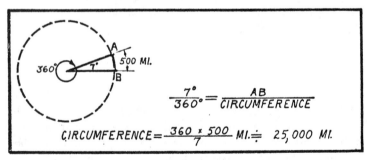

(b) *Calculation of the earth's circumference.*

is about 500 miles, so he deduced that the circumference of the earth is about 25,000 miles. This early value differs by less than 200 miles from the one now accepted.

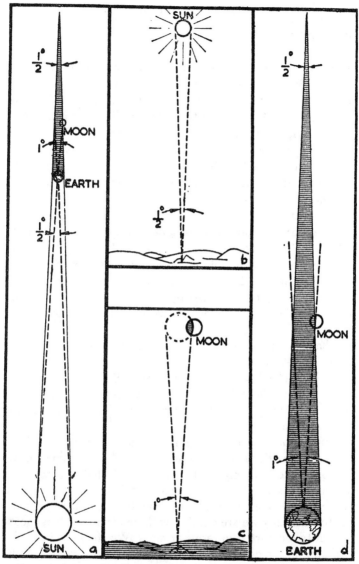

Fig. 2. Aristarchus' method of measuring the distance of the moon. (a) A partial eclipse of the moon. (b) The angular diameter of the sun as seen from the earth. (c) The angular diameter of the earth's shadow at the distance of the moon. (d) Construction for determining the moon's distance from the earth.

THE DISTANCE AND SIZE OF THE MOON. The first man to propose a method of measuring the distance to the moon appears to have been Aristarchus of Samos, who lived in the third century B.C. So slow was the development of science at that time, however, that his method was not used for another hundred and fifty years. It was then taken up by the great astronomer Hipparchus, who succeeded in obtaining a very accurate value. The method consists of the following steps (Fig. 2):

(i) Measurement of the angle subtended by the sun at the earth—that is, the sun's *angular* diameter (Fig. 2b). This angle is about ½ degree.

(ii) Measurement of the angular diameter of the earth's shadow at the distance of the moon (as observed at a partial eclipse of the moon). (Fig. 2c.) This angle is about 1 degree.

These observations suffice to determine the distance of the moon in terms of the earth's diameter. It can be calculated by trigonometry or measured by means of the following simple geometrical construction. Draw the shadow of the earth as shown in Fig. 2d. Since the sun is very distant, the edges of the shadow meet at nearly the same angle as that subtended by the sun—that is, ½ degree (Fig. 2a). Now draw in at the earth the angle of 1 degree determined in Step (ii). This angle must intersect the edge of the shadow at the position of the moon (Fig. 2d). We can then measure the distance of the moon in terms of the earth's diameter.

Hipparchus' observations imply that the distance of the moon is 30 earth diameters. Using Eratosthenes' value for the size of the earth, Hipparchus calculated that the moon is 240,000 miles away, a result which is within 1000 miles of the correct one.[1] We are here faced with a distance so much greater than any we are familiar with on earth that it is inconvenient to measure it in terms of miles. We shall use instead a modern unit, namely the time taken by light to travel the distance involved. Hipparchus' result implies that it takes light

[1] This refers to the average distance of the moon, since its actual distance varies by over 20,000 miles as it moves around the earth.

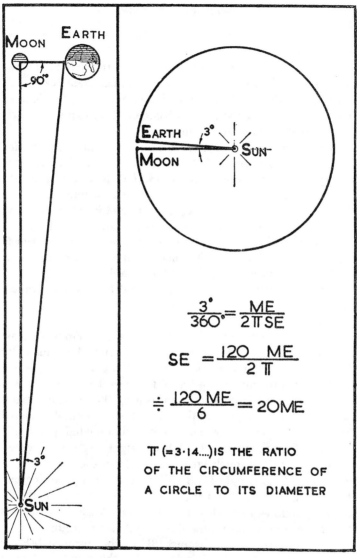

Fig. 3. Aristarchus' method of measuring the distance of the sun. (a) The earth-sun-moon system when the moon is half illuminated ("first quarter"). (b) Calculation of the sun's distance from the earth.

about 1¼ seconds to reach us from the moon—so the distance to the moon is 1¼ light-seconds.

Hipparchus was now in a position to determine the diameter of the moon. He simply measured its angular diameter, which is about ½ degree. The ratio of this angle to 360 degrees leads to a value for the moon's diameter of 2000 miles, which is a quarter of the earth's diameter.

In passing we may comment on a fact which the reader may have already noticed: that the moon's angular diameter is very nearly the same as the sun's. This is a very fortunate fact for astronomers, since it means that at a total eclipse of the sun the moon is just the right size to blot it out completely. When this happens we can see the outer extensions of the sun —its corona—which are too faint to be seen when the sun is shining (Plate I). Had the moon's angular diameter been smaller than the sun's we should not be able to see the corona at an eclipse, and solar astronomy would have lost an important source of information.

THE DISTANCE AND SIZE OF THE SUN. The first attempt to measure the distance to the sun also appears to have been made by Aristarchus. His method shows as much ingenuity as the one he suggested for measuring the distance to the moon, but it is far less accurate. It consists in measuring the angular separation between the moon and the sun at the moment of so-called "first quarter," when the moon is half illuminated (Fig. 3a).

In this figure the angle EMS is one right angle, 90 degrees. Since the sun S is much farther away from the earth than is the moon M, the angle MES will be very nearly 90 degrees. Aristarchus measured this angle and obtained the value 87 degrees. It follows from this that the angle MSE is 3 degrees (since the sum of the angles of a triangle is two right angles). We can now relate the distance SE to the distance ME, as shown in Fig. 3b. Aristarchus deduced in this way that the sun is about 20 times farther from the earth than is the moon. Since Hipparchus' value for the distance to the moon is 240,-000 miles, we see that according to Aristarchus' measurement

THE SIZE OF THE SOLAR SYSTEM

the distance to the sun is 4,800,000 miles, or 26 light-seconds.

From this estimate of the sun's distance we can also determine its diameter, since its *angular* diameter is ½ degree. In fact, since the angular diameter of the sun is the same as that of the moon, their diameters are in the same ratio as their distances (Fig. 4). According to Aristarchus, then, the sun's diameter is 40,000 miles.

Actually both these estimates are very bad approximations to the truth. The reason is that Aristarchus' method is very sensitive to an error in the measurement of the angle MES.

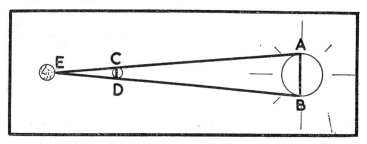

Fig. 4. Calculation of the sun's diameter. By similar triangles, the diameters of the sun and moon are in the same ratio as their distances.

For suppose that he had obtained a value of 89 degrees instead of 87 degrees—an error of only 2 per cent. In that case the angle ESM would be 1 degree. This is one third of the angle of 3 degrees actually obtained by Aristarchus, and it would imply that the sun is 60 times farther than the moon. Thus an error of only 2 per cent in the measured angle leads to an error of a factor 3 in the inferred distance to the sun. Aristarchus' error was in fact greater still, since the sun is actually about 400 times farther away than the moon. But this was not discovered till much later—Aristarchus' value was the accepted one for eighteen hundred years. Not until the eighteenth century was a better value obtained.

Modern Times

THE DISTANCE TO THE MOON. The first modern attempt to determine the distance to the moon was made by the amateur astronomer Abbé Nicholas Louis de la Caille (1713–62). He succeeded in measuring the *parallax* of the moon. Parallax is a very important concept in astronomy; it arises from the fact that an object looked at from different places appears to lie in

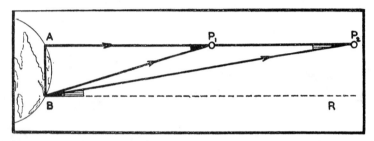

Fig. 5. Parallax. An object looked at from different places appears to lie in different directions. Here the parallax of P_1 relative to AB is the angle P_1BR. This is equal to the angle AP_1B, which is the angle subtended by the base line AB at the object P_1. Note that the parallax of the more distant object P_2 is less than that of P_1. In general the distance of P can be calculated from its direction, its parallax, and the length of the base line.

different directions. In Fig. 5 the angle P is known as the parallax of the object relative to the given base line AB; it is smaller the farther away the object is. Another way of describing it is to say that it is the angle subtended by the base line at the object. Its importance to astronomy arises from the fact that, if it can be measured relative to a base line of known length, the distance to the object can be calculated by simple trigonometry.

De la Caille was able to measure the moon's parallax relative to a base line stretching from Berlin to the Cape of Good Hope. He obtained a value of nearly 1 degree, which implies that the distance to the moon is about 60 earth radii, or 1¼ light-seconds. This result is very close to the one obtained by Hipparchus.

Both these results are indirect, in the sense that the moon's distance is measured in terms of the dimensions of the earth. Recent technical developments in radar have enabled a direct determination to be made. In 1957, B. S. Yaplee, an American radio astronomer, measured the time taken for an electromagnetic wave to reach the moon and be reflected back to the earth. So accurately can this time be measured that the distance to the moon is now known to within half a mile (corresponding to an accuracy of 2 parts in 1,000,000). The use of the light-second as the unit of distance is particularly appropriate in this case. At the time of writing, the moon is the most distant object from which radar echoes have been received; the next target is the planets.[2] This will call for equipment of greater power or sensitivity, but we have certainly not yet reached the limit of what is practicable.

THE DISTANCE TO THE SUN. Since the sun is much farther away than the moon, its parallax will be much smaller and is correspondingly more difficult to measure. It is therefore important to use the longest possible base line. In the eighteenth century various attempts were made, especially by Edmund Halley, to measure the parallax of the sun with the earth's diameter as base line. The actual method used was rather intricate and we need not go into the details here. The result for this parallax was about 18 seconds of arc.[3] This corresponds to a distance of 96 million miles, which is close to the modern value of 93 million miles. It is equivalent to 500 light-seconds or 8⅓ light-minutes.

Halley's result implies that Aristarchus underestimated the

[2] Echoes from Venus have now been detected.
[3] A degree is divided into 60 minutes of arc, and a minute into 60 seconds of arc.

distance to the sun by a factor of about 20. His value for the sun's diameter is too low by the same factor, of course; it should be about 800,000 miles, or 5 light-seconds.

THE DISTANCE TO THE PLANETS. The distances to most of the planets were also determined in the eighteenth century from measurements of their parallaxes. We shall not quote the individual distances to each planet, but to give a general idea of the size of the solar system, we may mention that the distance from the sun to the outermost known planet, Pluto (which was only discovered in 1930), is nearly 40 times the distance from the sun to the earth. In terms of light time this corresponds to 5½ light-hours—that is, it takes light 11 hours to cross the solar system.

Thus ends the first stage of our exploration of the universe. It has taken us from the circumference of the earth, a mere fraction of a light-second, to the extent of the solar system, a few light-hours. The next stage, to nearby stars, involves a big jump from light-hours to light-*years*. This jump is taken in the next chapter.

CHAPTER II

THE DISTANCE TO THE STARS

> If 25 years are required for a bullet out of a cannon, with its utmost swiftness, to travel from the sun to us, such a bullet would spend almost 700,000 years in its journey between us and the fixed stars. And yet when in a clear night we look upon them, we cannot think them above some few miles over our heads.
>
> CHRISTIAN HUYGENS

Introduction

To the problem of measuring stellar distances the Greeks contributed little or nothing. Some of them seem to have realized that the stars are very distant, but they were unable to measure any stellar parallaxes. This is not surprising, for the stars are so much farther away than the sun that their parallax is very small and difficult to detect, even when the base line is as long as possible—that is, when it is the diameter of the earth's orbit around the sun (Fig. 6). The fact that no stellar parallax had been detected by the seventeenth century even led some people to conclude that the earth did not move at all.

It was also in the seventeenth century that astronomers first made fairly accurate estimates of the distances to nearby stars, using physical considerations unknown to the Greeks. According to Christian Huygens, a famous Dutch scientist (1629–95), Sirius is about 30,000 times farther away than the sun. Huygens obtained this estimate in the following way. He looked at the sun through various small holes until he found

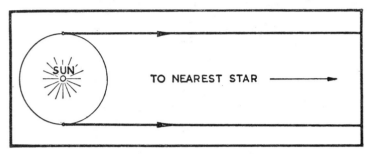

Fig. 6. Even the nearest stars are so far away that their parallax is very small.

one that reduced its brightness to that of Sirius. This enabled him to calculate how much farther away the sun would have to be to look like Sirius. He then assumed that the sun and Sirius are similar stars, and concluded that Sirius must be at just this distance.

Huygens' value of about 30,000 sun distances, or ½ light-year, is actually an underestimate, and a better value was ob-

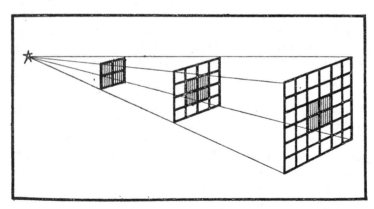

Fig. 7. The inverse square law for the apparent brightness of a star. The fraction of light intercepted by the shaded receiver decreases inversely as the square of its distance from the star.

tained by Newton (1642–1727). Newton used the inverse square law for the brightness of shining objects, a law which is illustrated in Fig. 7. Assuming as Huygens did that Sirius is as *intrinsically* bright as the sun, he obtained from his measurement of its apparent brightness a distance of 15 light-years. This result showed quite clearly why no stellar parallax had yet been observed, even when the base line was the diameter of the earth's orbit around the sun. For a distance of 15 light-years corresponds to a parallax of about ½ second of arc—this is roughly the angle subtended by a dime five miles away. The measurement of such a small angle was far beyond the techniques of the time.

Stellar Parallaxes

Another century passed before instrumental techniques developed to the point where angles of a fraction of a second could be measured. In the year 1838, F. W. Bessel, a German astronomer, (1784–1846), using a telescope of his own construction, succeeded in detecting and measuring the first stellar parallax.

Bessel used a slightly different version of the parallax effect from the one we have already described. Bessel's version is the one noticed by many car drivers when their passengers accuse them of driving too fast. To the passenger, the speedometer needle may appear to point to 65 miles per hour, whereas, when looked at face on by the driver, it points to only 55 miles per hour.[1] In the same way nearby stars shift in position relative to distant stars when viewed from different places (Fig. 8). When the parallax angle is small it is more accurate to measure it in relation to distant stars, rather than relative to a fixed direction. Ignoring the parallax of the distant stars will introduce a smaller error than attempting to determine a fixed direction from two widely separated places.

[1] This assumes that the car has a right-hand drive.

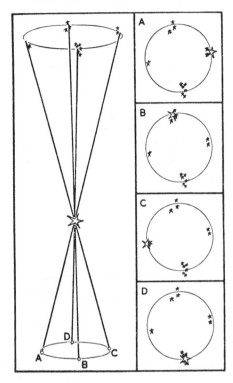

Fig. 8. Parallax of a nearby star relative to distant stars. When viewed from different positions in the earth's orbit, the nearby star appears to move relative to distant stars.

The first problem facing Bessel was to select a star with as large a parallax as possible. This meant selecting one of the nearest stars, but how could he tell which stars are nearby before any distances to stars had been measured? The most obvious answer was to use Newton's criterion of a large apparent brightness. However, since stars might vary greatly in their intrinsic brightness, Bessel preferred to use another criterion, arising from the fact that the so-called "fixed" stars are actually in motion relative to one another (Fig. 9a).

THE DISTANCE TO THE STARS

This motion is of two kinds: radial motion, which is along the line of sight (Fig. 9b), and transverse motion, which is at right angles to the line of sight (Fig. 9c). Only the transverse motion produces a change in the position of a star as viewed from the earth. In particular, if two stars have the same transverse velocity, the nearer star will move through a larger angle than the farther star in the same interval of time (Fig. 10). As a result the relative positions of the stars in the sky will change. This effect is familiar to people who have watched airplanes in the sky; distant airplanes appear to be almost stationary because they move through a tiny angle during the time they are being watched, whereas a nearby airplane will move through an appreciable angle in quite a short time.

Now Halley had discovered in 1718 that several stars were no longer in the position recorded long before by Hipparchus and Ptolemy; as a result of their transverse velocity they had moved relative to other stars by as much as the moon's angular diameter—that is about ½ degree. The angle they moved through in a year Halley called their "proper motion." By Bessel's time the proper motions of many stars had been measured, and he decided that a large proper motion was a better guide than brightness to a star's proximity.

Now in Bessel's time one particular star—61 Cygni—was famous for its large proper motion. It moves relative to other stars through 5 seconds of arc in a year, which by astronomical standards is considerable—indeed, it was known at the time as the "flying star." Bessel therefore decided to concentrate his attention on this star and after twelve months of careful observations he succeeded in obtaining in December 1838 a parallax of 0.31 seconds of arc. The base line for this parallax was the diameter of the earth's orbit around the sun, so Bessel deduced that the star is 11 light-years away. With this result man had at last succeeded in measuring distances beyond the solar system.

As so often happens in science, Bessel's great achievement was not an isolated triumph. Only two months after he had announced his result the British astronomer Henderson suc-

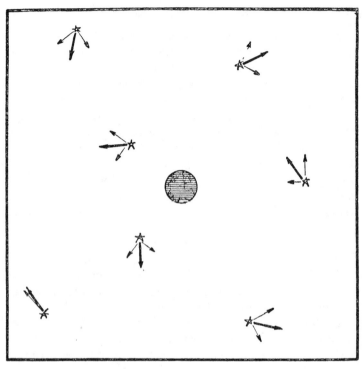

Fig. 9. (a) Moving stars. Their radial and transverse velocities are indicated.

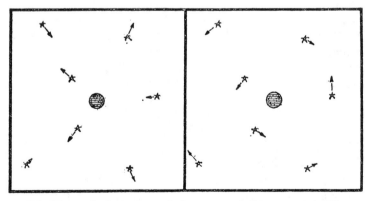

(b) The radial motion of the stars of Fig. 9a. (c) The transverse motion of the stars of Fig. 9a.

ceeded in measuring the parallax of another star, Alpha Centauri, which he placed much nearer than 61 Cygni—namely, 4 light-years away. Although modern measurements have increased this value by 30 per cent, Alpha Centauri is still the nearest known star. Yet another stellar parallax was measured two years later, in 1840, by the Russian astronomer Struve (the first of a long line of famous astronomers that survives to the present day). Struve succeeded in measuring the parallax of the star Vega, although it is about three times farther away than 61 Cygni.

The excitement which these three measurements created at the time is evident in the remarks of Sir John Herschel when, as president of the Royal Astronomical Society, he awarded the society's Gold Medal to Bessel in 1841.

> I congratulate you and myself that we have lived to see the great and hitherto impassable barrier to our excursions into the sidereal universe—that barrier against which we have chafed so long and so vainly—almost simultaneously overleaped at three different points. It is the greatest and most glorious triumph which practical astronomy has ever witnessed. Perhaps I ought not to speak so strongly—perhaps I should hold some reserve in favour of the bare possibility that it may all be an illusion and that further researches, as they have repeatedly before, so may now fail to substantiate this noble result. But I confess myself unequal to such prudence under such excitement. Let us rather accept the joyful omens of the time and trust that, as the barrier has begun to yield, it will speedily be prostrated. Such results are among the fairest flowers of civilization.

However, Herschel's optimism was premature; the barrier was prostrated only sixty years later when the photographic plate replaced the eye at the end of the telescope. This enabled systematic programs of parallax measurement to be carried out. The resulting increase of knowledge was tremendous. In 1900, when the photographic method was introduced, the dis-

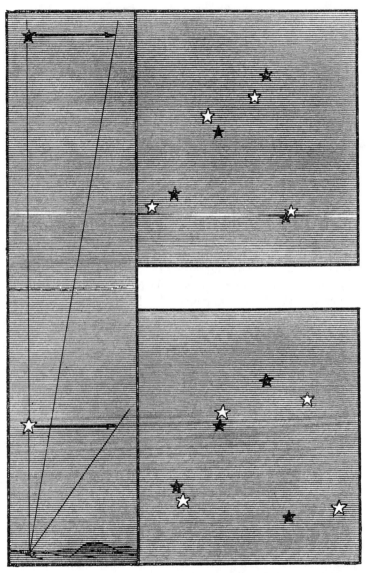

Fig. 10. *The proper motions of near and distant stars. (a) A nearby star moves through a larger angle in unit time than does a distant star which has the same transverse velocity. The nearer star is then said to have a larger*

tances to only 60 stars were known; by the end of 1957 the number of known distances had increased to over 10,000.

The errors in the best-established parallaxes are about one hundredth of a second of arc, which corresponds to the angle subtended by a pinhead at ten miles. This is probably near the limit of attainable accuracy, which means that stars more than a few hundred light-years away have parallaxes which are too small to measure. We here meet a new barrier; can it too be prostrated?

Indirect Methods of Measuring Distance

One can still measure the apparent brightness of a star whose parallax is too small to be detected. This suggests that we might revive Newton's method of deducing distances from the inverse square law, by assuming, as he did, that the distant stars have the same intrinsic brightness as the sun. Unfortunately, as Bessel feared, Newton's method is unreliable. The parallax measurements show that stars vary greatly in their intrinsic brightness, some stars, for instance, being 50,000 times brighter than the sun. Nevertheless, if the intrinsic brightness of a star could be determined by some indirect method, a measurement of its apparent brightness would still enable us to determine its distance. So what we seek is some measurable property of a star that is correlated in a unique way with its intrinsic brightness.

Such a property was discovered by Adams and Kohlschutter in 1914. They had been studying the light emitted by stars after it had passed through an instrument known as a spectroscope, which separates out into a spectrum the different wave

proper motion. (b) Near (white) and distant (black) stars at two different times. Only the near stars show a measurable change of position in the sky. One can thus identify near stars, which are the only ones whose parallaxes can be measured.

lengths or colors present in light. The spectrum of a star consists of a series of separate dark lines superimposed on a continuous band of light (Plate II). The positions of the lines in the spectrum depend on the nature of the atoms present in the star, atoms of particular elements giving rise to characteristic patterns of lines. The darkness of the lines, on the other hand, depends on the number of atoms which form them and also on the physical conditions in their environment.[2]

Adams and Kohlschutter found that some lines are darker than others by an amount which varies from star to star. Now they also knew the absolute brightness of those stars whose parallax had been measured. They were therefore able to investigate whether the relative darknesses of the spectral lines were correlated with the absolute brightness of a star. They did, in fact, find such a relation. This enabled them to infer the intrinsic brightness of a star simply from measurements of the darkness of its spectral lines, without its parallax having to be known. This is just what is needed for measuring the distance of a star too far away for its parallax to be detectable. The range of distance measurement was extended in this way from hundreds of light-years out to thousands of light-years, where the spectrum of a star can still be photographed.

This range of distance was large enough for the beginnings of *organization* to be discernible in the arrangement of the stars in space. In order to explore this organization more fully, however, the range of measurable distances had to be further extended. Fortunately it was discovered that the stars known as Cepheid variables could be used as distance indicators even at distances so large that spectra are too faint to be measured. The first such star was discovered in 1784, when the amateur astronomer John Goodricke found that the brightness of Delta Cephei fluctuated in a regular way every 5½ days (Fig. 11).

---

[2] Laboratory spectra show bright lines rather than dark ones. The difference arises because in the case of stars the atoms on the surface *absorb* light coming from the deep interior. Since an atom absorbs and emits light at exactly the same wave lengths, this difference does not affect the positions of the lines.

Many similar stars have since been discovered, and the generic name "Cepheid variable" has been given to them all. About a dozen of them can be seen with the naked eye, the best known being the Pole Star, which has a period of nearly 4 days.

In 1912, Miss Henrietta Leavitt discovered that there is a well-defined relation between the period of a Cepheid variable and its intrinsic brightness, the brightest stars having the longest periods (Fig. 12). This period-luminosity relation, as it is

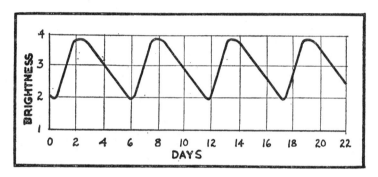

Fig. 11. The variation in brightness of the star Delta Cephei.

called, enables one to determine the intrinsic brightness of any Cepheid whose period can be measured. Its distance can then be determined from its apparent brightness. This method works for much greater distances than does the one using spectral lines, since a Cepheid's period can be measured even when it is too faint to give a good spectrum. The method was first used in 1913, to determine the distance to the Lesser Magellanic Cloud, one of two diffuse-looking objects which can be seen by the naked eye in the southern hemisphere (Plate III) and which contains many Cepheids. The result obtained was about 100,000 light-years, which was by far the greatest distance that had then been measured.

Once distances as great as 100,000 light-years could be measured, the organization of the stars in space was readily

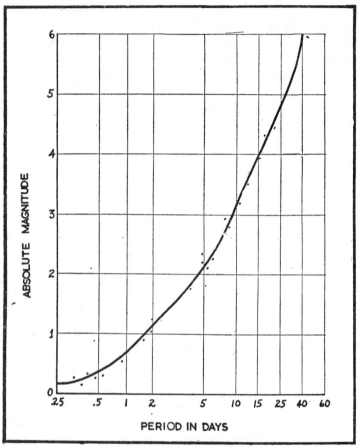

Fig. 12. The period-luminosity relation for Cepheid variables. The larger the intrinsic brightness of a Cepheid, the longer its period. (The conventional astronomical scale of brightness is the negative of that used here.) A measurement of its period thus enables its intrinsic brightness to be deduced. Its distance can then be inferred from its apparent brightness.

established. In the hundred and fifty years leading up to the discovery of the period-luminosity relation for Cepheids, there had been much speculation about this organization. Particular attention had been paid to the structure of the Milky Way (Plate V), the band of light that can be seen on a clear night stretching across the sky. Galileo had discovered that this band of light actually comes from a vast collection of faint stars which the naked eye cannot distinguish. The sun itself and the nearby stars also appear to belong to this collection, and the name "Milky Way" has come to be used for all the stars associated with the sun rather than for the appearance which gave rise to it. After centuries of work the size and shape of this collection of stars were definitely established in 1952. This achievement is described in the next chapter.

CHAPTER III

THE MILKY WAY

> Pricked out with less and greater lights, between the poles of the universe, the Milky Way so gleameth white as to set very sages questioning.
> DANTE

Eighteenth-Century Speculations

Whoever turns his eye to the starry heavens on a clear night, will perceive that streak or band of light which on account of the multitude of stars that are accumulated there more than elsewhere, and by their getting perceptibly lost in the great distance, presents a uniform light which has been designated by the name *Milky Way*. It is astonishing that the observers of the heavens have not long since been moved by the character of this perceptibly distinctive zone in the heavens, to deduce from it special determinations regarding the position and distribution of the fixed stars. For it is seen to occupy the direction of a great circle, and to pass in uninterrupted connection around the whole heavens: two conditions which imply such a precise destination and present marks so perceptibly different from the indefiniteness of chance, that attentive astronomers ought to have been thereby led, as a matter of course, to seek carefully for the explanation of such a phenomenon.

So wrote the famous philosopher Immanuel Kant in 1755. It was indeed only in his own time that detailed speculations

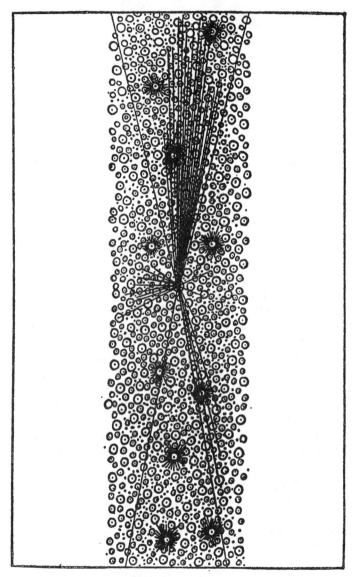

Fig. 13. Thomas Wright's model of the Milky Way.

were made about the significance of the Milky Way. In particular, three eighteenth-century speculators came very near the truth. They were Thomas Wright of Durham (1711–86), Kant himself (1742–1804), and Johann Lambert (1728–77).

The first of these pioneers was Thomas Wright. He suggested that the Milky Way consisted of a flattened distribution of stars forming a disc. In 1750 he wrote as follows:

> This is the great order of nature which I shall now endeavour to prove and thereby solve the Phaenomena of the *Via Lactea*.
>
> Let us suppose the whole frame of nature in the form of an artificial horizon of a globe, I do not mean to affirm that it really is so in fact, but only state the question thus to help your imagination to conceive more aptly what I would explain. (Fig. 13) will then represent a just section of it. Now in this space let us imagine all the Stars scattered promiscuously, but at such an adjusted distance from one another, as to fill up the whole medium with a kind of regular irregularity of objects. And next let us consider what the consequence would be to an eye situated near the centre point, or any where about the middle plane, as at the point A. Is it not, think you, very evident, that the Stars would there appear promiscuously dispersed on each side, and more and more inclining to disorder, as the observer would advance his station towards either surface. If your optics fails you before you arrive at these external regions, only imagine how infinitely greater the number of Stars would be in those remote parts, arising thus from their continual crowding behind one another; thus, all their rays at last so near uniting, must meeting in the eye appear, as almost in contact, and form a perfect zone of light; this I take to be the real case, and the true nature of our *Milky Way*.

These speculations were carried further by Kant in 1755:

> It was reserved for an Englishman, Mr. Wright of Durham, to make a happy step with a remark which does

not seem to have been used by himself for any very important purpose, and the useful application of which he has not sufficiently observed. He regarded the Fixed Stars not as a mere swarm scattered without order and without design, but found a systematic constitution in the whole universe. We would attempt to improve the thought which he thus indicated, and to give to it that modification by which it may become fruitful in important consequences whose complete verification is reserved for future times. . . .

I come now to that part of my theory which gives it its greatest charm, by the sublime idea which it presents of the plan of the creation. The train of thought which has led me to it is short and natural; it consists of the following ideas. If a system of fixed stars which are related in their positions to a common plane, as we have delineated the Milky Way to be, be so far removed from us that the individual stars of which it consists are no longer sensibly distinguishable even by the telescope; if its distance has the same ratio to the distance of the stars of the Milky Way as that of the latter has to the distance of the sun; in short, if such a world of fixed stars is beheld at such an immense distance from the eye of the spectator situated outside of it, then this world will appear under a small angle as a patch of space whose figure will be circular if its plane is presented directly to the eye, and elliptical if it is seen from the side or obliquely. The feebleness of its light, its figure, and the apparent size of its diameter will clearly distinguish such a phenomenon when it is presented, from all the stars that are seen single.

We do not need to look long for this phenomenon amongst the observations of the astronomers. It has been distinctly perceived by different observers. They have been astonished at its strangeness; and it has given occasion for conjectures, sometimes to strange hypotheses, and at other times to probable conceptions which, however, were just as groundless as the former. It is the "nebulous" stars which we refer to, or rather a species of them, which M. de Maupertuis thus describes: "they are," he says, "small luminous patches, only a little more brilliant than the dark background of the heavens; they are presented in all quarters; they present the figure of ellipses more or less open; and their light is much feebler than that of any other object we can perceive in the heavens."

It is natural and conceivable to regard them as being not enormous single stars but systems of many stars, whose distance presents them in such a narrow space that the light which is individually imperceptible from each of them, reaches us, on account of their immense multitude, in a uniform pale glimmer. Their analogy with the stellar system in which we find ourselves, their shape, which is just what it ought to be according to our theory, the feebleness of their light: all this is in perfect harmony with the view that these elliptical figures are just universes and, so to speak, Milky Ways. And if conjectures, with which analogy and observation perfectly agree in supporting each other, have the same value as formal proofs, then the certainty of these systems must be regarded as established.

The same idea was proposed by Lambert in 1761. In his *Cosmological Letters* he wrote:

It would not be at all astonishing then, if milky ways, situated far from us in the depths of the heavens, should make no impression on the eye whatever. But who knows whether the pale light that is observed in Orion, is not one of the milky ways nearer to us than the rest? And, perhaps, by diligent application of the telescope, we may discover elsewhere similar appearances. But where is the milky way itself in relation to other milky ways? Here ends all our science with the utmost stretch of our eyes and instruments.

These quotations show that by the middle of the eighteenth century, even before the scale of stellar distances had been established, various thinkers had concluded that the stars were organized in a definite disclike system, and that other stellar systems, identified with the "nebulous stars" or nebulae, lay outside. These ideas remained speculative until the end of the eighteenth century, when systematic star counts, particularly by Sir William Herschel, the father of modern stellar astronomy, confirmed some of their main features.

The Herschel-Kapteyn Model of the Milky Way

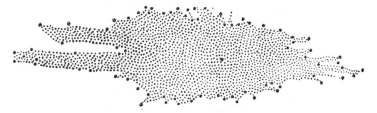

Fig. 14. Sir William Herschel's model of the Milky Way.

Sir William Herschel (1738–1822) described his first results in a paper called "On the Construction of the Heavens," published in 1785. His counts of stars lying in different directions, conducted over many years, led him to adopt the model of the Milky Way proposed by Wright, Kant and Lambert (Fig. 14). In addition Herschel followed Kant and Lambert in thinking that the nebulae (many more of which had been discovered by Herschel himself) are probably external Milky Ways or galaxies.[1]

Herschel's work on star counts was continued by his son, Sir John Herschel (1792–1871), who went south to observe those regions of the sky which are not visible in the north. In 1847 he published his observations of 1700 nebulae and 70,000 stars. His analysis of this mass of material confirmed

---

[1] "Galaxias" is the Greek for "milk."

his father's disclike model and dominated astronomical thinking about the Milky Way for the next sixty years.

The problem was reopened in 1906 by the Dutch astronomer J. C. Kapteyn (1851–1922), who suggested that many observatories should collaborate in a detailed study of 206

Shapley's Model of the Milky Way

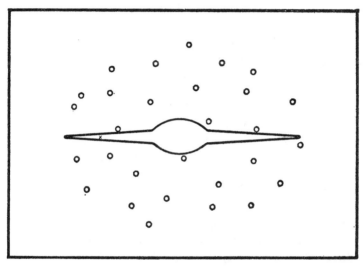

Fig. 15. Shapley's model of the Milky Way. Each circle represents a globular cluster. The hub and disc contain most of the stars in the Milky Way. The sun, which is two thirds of the way out from the center, is marked by a cross.

selected areas of the sky. His suggestion was acted upon and after analyzing all the results he was able to publish his definitive model of the Milky Way in 1922. This had the same general disclike structure as the Herschel model but was much more detailed, particularly of course in its scale, since the Herschels had had to work with virtually no distance determinations. Kapteyn believed the Milky Way to be a circular disc 27,000 light-years across and 5400 light-years thick, in the center of which he placed the sun. Despite the tremendous amount of information that went into the construction of Kapteyn's model, however, we now know that it is incorrect in several important respects. These errors were discovered mainly through the use of the Cepheid variable method of determining distances, the results of which led Harlow Shapley to propose a new model. This model, with minor modifications, is the one accepted today.

We have seen how powerful is the Cepheid variable method for determining large distances. In the hands of the American astronomer Harlow Shapley it dethroned the sun from its central position in the Herschel-Kapteyn model of the Milky Way. Shapley made his discovery while studying globular clusters—these are symmetrical star clusters which contain from 10,000 to 1,000,000 stars (Plate VI). With the aid of the 60-inch telescope on Mount Wilson, which was then the largest in the world, Shapley used the Cepheid variable method to determine the distances of about 100 globular clusters. From these distances he derived their distribution in space.

In 1918 he announced his results. According to Shapley, the globular clusters are arranged in a slightly flattened system whose center is about 50,000 light-years from the sun in the direction of Sagittarius (Plate VII). Shapley suggested that the center of this system of globular clusters is also the center of our galaxy, and that their extents are the same (about 300,000 light-years) (Fig. 15). The fact that the sun is well away from the center of the system of globular clusters is obvious from the fact that most globular clusters lie on one side of the sun. But it is not obvious that the system of globular clusters is closely related to the structure of the Milky Way as a whole. The conflict between the Shapley and Herschel-Kapteyn models, particularly as regards the position of the sun, could not be settled without further evidence. This came shortly afterwards with two developments which resolved the conflict in favor of Shapley; to be more precise, one development showed that Shapley was right and the other showed that Kapteyn was wrong.

The first development occurred in 1926–27, when Lindblad and Oort discovered from the motions of the stars that the

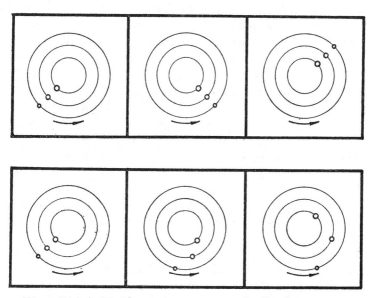

Fig. 16. (a) Rigid rotation, as in a wheel. Three points originally on a straight line remain on a straight line. (b) Differential rotation, as in the solar system. Points near the center move ahead of those farther out.

Milky Way is rotating. This rotation had been previously postulated by Kant in order to explain the flattened form of the Milky Way. Furthermore Oort found, as Kant had suggested, that the Milky Way does not rotate like a rigid wheel but rather in the same way as the planets rotate about the sun (Fig. 16). The difference between these two types of rotation lies in the fact that, whereas in a rigid wheel every particle rotates with the same period, in the solar system the nearer a planet is to the sun the shorter is its period. This is a characteristic property of circular motion around a central attracting mass. It is exemplified by the motion of artificial satellites around the earth. The first and nearest satellite had a period of 95 minutes, whereas the second had a period of 103 minutes. In the same way the stars in our galaxy have a shorter period of rotation the nearer they are to its center (where most of the mass of the galaxy is concentrated).

It is fortunate that the stars move in this way since, if they all had the same period, their rotation would be very difficult to detect. The reason for this is that all the stars would then remain the same distance apart, so that their rotation would give them neither proper motion nor radial velocity relative to the sun (Fig. 17d). On the other hand, if they had any radial velocity it could be detected by the so-called Doppler effect (as explained on p. 56). Fortunately the stars do rotate with different periods, so that their distances are changing and they have detectable radial velocities. The pattern of *total* velocities relative to the sun is similar to that observed by drivers on a multiple-lane highway. The central lane of a stream of traffic will carry cars of intermediate speed; relative to these cars, those on the inner lane move forward and those on the outer lane move back. So it is with the sun; the stars nearer the center of rotation move forward and those farther away move back. The resulting pattern of total and radial velocities is shown in Fig. 17b and c, for stars near the sun.

The existence of this pattern of radial velocities was established by Oort in 1927. He was able to deduce from his observations that the sun makes a complete revolution in about 200 million years. The center about which this revolution takes place is in the direction of Sagittarius, in agreement with Shapley's suggestion that the center of the Milky Way lies in this direction. The agreement was not complete, however, since according to Oort the center is only 30,000 light-years away, instead of the 50,000 light-years obtained by Shapley. Apart from this quantitative discrepancy, which we shall clear up shortly, Oort's work confirmed Shapley's suggestion that the

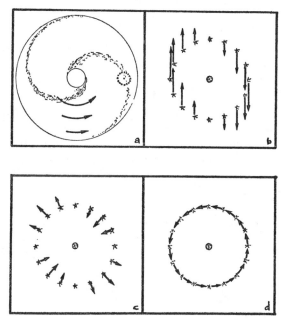

Fig. 17. (a) The differential rotation of the Milky Way. The stars near the center move faster than those farther away. The sun and its neighboring stars (on the right of the diagram) move at different speeds, as shown in b. (b) The differential motion of the stars relative to the sun. The stars nearer the center of the Milky Way move ahead, while those farther away drop back. (c) The radial motions of the stars relative to the sun. These radial motions have been detected and the differential rotation of the Milky Way thereby inferred. (d) The motions of the stars relative to the sun, if the Milky Way rotated like a rigid wheel. These motions are entirely transverse and would be very difficult to detect since they do not lead to a change in the relative positions of the stars in the sky.

sun is not in a privileged position at the center of the Milky Way but is an undistinguished star some way out.

The second development, which showed directly how Kapteyn's model was wrong, was the discovery that the stars are dimmed by a fog which permeates the space between them. So thick is this fog that it completely concealed most of the Milky Way from Kapteyn, but, as we shall see, only partially from Shapley. The fog was discovered as a by-product of investigations on the so-called "open" clusters of stars. These differ from the globular clusters studied by Shapley in that they contain fewer stars and are much less compact and symmetrical (Plate VIII).

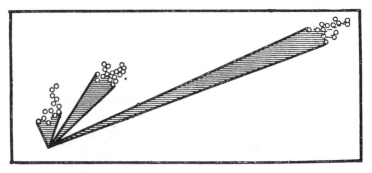

Fig. 18. The angular diameters of open clusters of stars. If the clusters have equal actual diameters their distances are inversely proportional to their angular diameters.

In 1930 the Swiss astronomer R. J. Trumpler compared two different methods of determining the distances to these open clusters. The first method is to measure their apparent brightness and to assume that they all have the same intrinsic brightness. This method is very unreliable for stars because of their great variation in intrinsic brightness. Fortunately it is more reliable for clusters of stars, since the differences between the various stars in a cluster will tend to average out.

The second method used by Trumpler was to measure the

angular diameters of the clusters (Fig. 18). He here had to assume that all the clusters have the same dimensions. The distance of a cluster is then inversely proportional to its angular diameter.

Trumpler found that these two methods gave different results for the distances to his open clusters. The most obvious explanation of this discrepancy is that open clusters are not in fact intrinsically similar. But this would not account for Trumpler's further discovery that the discrepancy is systematically greater the fainter the clusters. He was able to explain this only by postulating the existence of an interstellar fog which absorbs light from the clusters. The dimming effect of this fog makes the clusters seem farther away than they really are when their distance is judged by their apparent brightness, but it has no effect, of course, on their angular diameters. Furthermore, the resulting error will be greater for the more distant clusters, whose light is more heavily absorbed. This neatly explains the systematic behavior of the discrepancy.

When a correction is made for the fog the results of the two methods can be reconciled. Trumpler had to assume that the fog lies near the plane of the galaxy and is about 600 light-years thick (Fig. 19). The existence of such a fog had been suspected earlier, but Trumpler's evidence was the most convincing. Later work has amply confirmed it. To describe the details of this work would take us too far afield, but we may notice in passing that the existence of this fog gives an immediate explanation of an effect noticed by Sir John Herschel, namely that nebulae seem to avoid the plane of the galaxy and to exist mostly well away from it, where the fog is fairly thin.

The discovery that light from the stars is absorbed on its way to the earth showed that Kapteyn had a misleading idea of how deeply into the Milky Way his observations had penetrated. Clearly his observations related to stars that lie nearer to the sun than he supposed—stars which in fact are nowhere near the edges of the Milky Way. This explains why he constructed a model of the Milky Way with the sun at its center.

Shapley, on the other hand, had observed many globular clusters lying well above the plane of the galaxy, where the fog is less effective. This gave him a more complete view of the galaxy. Nevertheless, the fog still influenced some of Shapley's numerical results, especially those that refer to globular clusters lying in or near the plane of the galaxy. As a result his value for the distance to the center of the system of globular

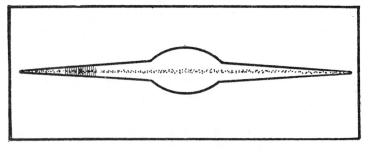

Fig. 19. The belt of fog in the Milky Way. The shaded portion indicates the region explored by Kapteyn, which he mistook for the whole of the Milky Way.

clusters should be reduced from 50,000 to about 30,000 light-years. This just eliminates the discrepancy with Oort's value deduced from the rotation of the Milky Way. In addition the diameter of the Milky Way must be reduced even more—from 300,000 light-years to about 100,000 light-years. This illustrates how the nuisance value of the fog increases with increasing distance. Indeed, the center of the galaxy, which is believed to be extremely bright intrinsically is practically invisible, only about one hundredth of 1 per cent of its light being transmitted to us.

THE MILKY WAY

The Milky Way as a Spiral Galaxy

Once the main features of the Milky Way's structure were established, astronomers turned their attention to its more detailed properties. In particular, they tried to discover whether the Milky Way is a *spiral* galaxy. The point of this investigation is as follows. Back in the mid-nineteenth century the amateur astronomer Lord Rosse (1800–67) had built an enormous telescope and discovered to the amazement of his contemporaries that several of the nebulae looked like spirals (Plate X). If Kant were right and these were external Milky Ways, the intriguing possibility arose that our own galaxy is also a spiral.

In fact many features of the distribution of stars suggested fairly strongly that our galaxy is indeed a spiral, and in 1952 the American astronomer W. W. Morgan showed conclusively that very bright stars (supergiants) lie along spiral arms, one of which passes just by the sun. From outside, our galaxy looks rather like the Andromeda nebula (Plate IX). This great discovery was the culmination of many years' work, which two decades ago was thought to be impossibly difficult. Nevertheless, in the same year came dramatic and unequivocal confirmation of the spiral nature of our galaxy from the use of a completely new technique whose power we are only beginning to explore.

In 1944 the Dutch astronomer H. C. Van de Hulst recalled that the spectrum (see p. 23) of hydrogen contains a line whose wave length is 21 centimeters—that is, a wave length lying in the *radio* region of the spectrum, in contrast to the familiar spectral lines which lie in the visible region (Fig. 20). Now it was believed that clouds of hydrogen gas exist in interstellar space although these had not yet been detected. These clouds would emit a 21-centimeter line which is very weak and hard to observe, but electronic techniques of ampli-

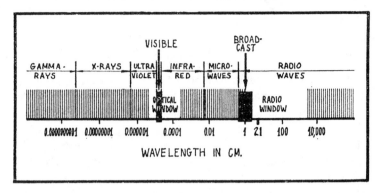

Fig. 20. The electromagnetic spectrum. The two windows refer to the wave length regions for which the atmosphere is transparent.

fication have become so powerful that Van de Hulst suggested that they could be used to detect it. This fruitful suggestion was successfully verified in 1951 by Ewen and Purcell at Harvard, Oort and Müller at Leiden, and Christiansen and Hindman in Australia; interstellar hydrogen clouds had at last been detected. Only one year later the Dutch astronomers were able to draw a preliminary map of the interstellar hydrogen in the Milky Way, showing clearly that it was mostly confined to spiral arms. Furthermore, these spiral arms agreed well in position with those obtained by Morgan. This was very gratifying, since a close association between interstellar hydrogen and supergiant stars had been suspected for some time.

These 21-centimeter results were announced at the 1952 meeting of the International Astronomical Union in Rome and created a great impression. Subsequent studies have confirmed and amplified them (Fig. 21). Nevertheless, despite these remarkable results, Oort always emphasizes that we are still only in the Herschel phase of 21-centimeter astronomy. Tremendous new discoveries may confidently be expected.

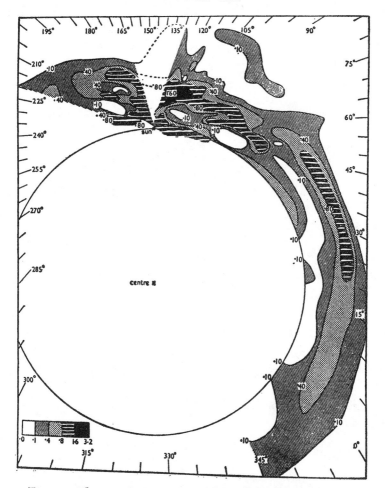

Fig. 21. The spiral arms of the Milky Way as revealed by 21-centimeter radiation. The arms in this diagram consist of clouds of hydrogen rather than the stars which can be seen in spiral galaxies.

CHAPTER IV

EXTERNAL GALAXIES

> I don't pretend to understand the Universe—
> it's a great deal bigger than I am.
> CARLYLE

Introduction

We must now face the fundamental question: Does the Milky Way comprise the total contents of the universe or was Kant correct in supposing that the universe contains many other galaxies much like our own? Kant had gone further and suggested that the external galaxies were the objects known to astronomers as nebulae. However, since Kant's time it had been discovered that these nebulae were of several kinds: in particular, astronomers distinguished between those that could be resolved into stars by powerful telescopes and those which seemed to be luminous gas. Of the resolved nebulae some were globular clusters of stars while others had a spiral appearance.

A great controversy arose as to whether the spiral nebulae were inside or outside the Milky Way. Many such nebulae had been discovered since Lord Rosse first catalogued 14 of them. In particular, the use of photographic plates with long exposures showed that there existed many more. Even before the 100-inch telescope came into use in 1918 it had been estimated that there exist at least 500,000 spiral nebulae. In order to illustrate the state of opinion on the whereabouts of these nebulae we shall quote two conflicting comments. The first comes from a historian of astronomy, Agnes Clarke, writing in 1905:

> The question whether nebulae are external galaxies hardly any longer needs discussion. It has been answered by the progress of research. No competent thinker, with the whole of the available evidence before him, can now, it is safe to say, maintain any single nebula to be a star system of co-ordinate rank with the Milky Way.

The main evidence for this view was the fact that the distribution of nebulae is closely related to the structure of the Milky Way—all the nebulae lie in directions well away from its disc. We now know, however, that this feature of their distribution is only apparent. It is caused by the fog that lies in the disc, an explanation already proposed by the English astronomer, Sir Arthur Eddington in 1914, sixteen years before Trumpler established that the fog really exists:

> In the days before the spectroscope had enabled us to discriminate between different kinds of nebulae, when all classes were looked upon as unresolved star clusters, the opinion was once widely held that these nebulae were "island universes," separated from our own stellar system by a vast empty space. It is now known that the irregular gaseous nebulae, such as that of Orion, are intimately related with the stars, and belong to our own system; but the hypothesis has recently been revived so far as regards the spiral nebulae. . . .
>
> It must be admitted that direct evidence is entirely lacking as to whether these bodies are within or without the stellar system. Their distribution, so different from that of all other objects, may be considered to show that they have no unity with the rest. Indeed, the mere fact that spiral nebulae shun the galaxy may indicate that they are influenced by it. The alternative view is that, lying altogether outside our system, those that happen to be in low galactic latitudes are blotted out by great tracts of absorb-

ing matter similar to those which form the dark spaces of the Milky Way.

If, however, it is assumed that these nebulae are external to the stellar system, that they are in fact systems co-equal with our own, we have at least an hypothesis which can be followed up, and may throw some light on the problems that have been before us. For this reason the "island universe" theory is much to be preferred as a working hypothesis; and its consequences are so helpful as to suggest a distinct probability of its truth.

The first substantial piece of evidence that Eddington was correct came from the work of the American astronomer H. D. Curtis (1872–1942), who discovered in 1918 that the great spiral nebula in Andromeda contained stars whose brightness suddenly flares up and slowly dies away again. Curtis assumed that these novae, as they are called, have the same intrinsic brightness as the ones already discovered in the Milky Way. He then measured their apparent brightness and deduced the distance to the Andromeda nebula. His result was about 1,000,000 light-years, a distance so much greater than the size of the Milky Way as to indicate that the Andromeda nebula is indeed outside our galaxy.

The Hubble Era (1924–36)

Curtis' result was confirmed in 1924 by the American astronomer Edwin P. Hubble (1889–1953). Using the 100-inch telescope, Hubble found that there were Cepheid variables in the Andromeda nebula and also in other spirals. He was then able to determine their distance by using the period-luminosity relation (p. 25). In this way he obtained a distance of 800,000 light-years for the Andromeda nebula, and similar values for other spiral nebulae. Now that these nebulae are established as stellar systems outside our own, we shall henceforth call them galaxies. In the years that followed Hubble studied these galaxies in great detail and in 1936 he summarized this work in his book *The Realm of the Nebulae.*

One of Hubble's main achievements was to establish methods of measuring the distances to galaxies which are too far away for their Cepheid variables to be detected. He did this in a series of steps, starting with the measurement of distances by means of Cepheid variables. This first step works only for the nearest galaxies, which are clustered together into what is known as the "local group," to which the Milky Way belongs.

For his next step Hubble used supergiant stars, which are intrinsically brighter than Cepheids. These stars can be detected in many distant galaxies whose Cepheids are invisible. Hubble assumed that the brightest supergiants in all the galaxies have about the same intrinsic brightness; their apparent brightness then indicates their distance. Hubble checked this method by using it on the local group, whose distances were already known from their Cepheid variables, and then applied it to more distant galaxies. In this way he extended the range of measurable distances from 1 million light-years to 10 million light-years.

This was as far as Hubble could go using stars as distance indicators. All that was left for him in the final step was to turn to galaxies themselves, and to assume that they all have the same intrinsic brightness. That this step is a reasonable one was indicated by his observations of a large cluster of galaxies in the direction of the Virgo constellation—a cluster that he estimated to be about 8 million light-years away. His measurements of the apparent brightnesses of galaxies in this cluster showed that their intrinsic brightnesses differ from one another by at most a factor of 10. The assumption that all galaxies have an intrinsic brightness in the middle of this range of values would then lead to an error of about a factor 3 in unfavorable cases. The total range of distances is fortunately so large that this uncertainty will not conceal any systematic trend in the properties of galaxies at different distances.

By using these methods, Hubble explored the universe out to the fantastic distance of 500 million light-years—a region

EXTERNAL GALAXIES

which contains about 100 million galaxies. This is a considerably greater number than the 500,000 which were known before the days of the 100-inch telescope. Hubble's book contains many details about some of these galaxies, but we can only describe his main results (some of which have had to be considerably modified in recent years). According to Hubble, the average distance between galaxies is about a million light-years. They are, on the whole, much smaller than the Milky Way, which has a diameter of about 100,000 light-years. In fact the majority of galaxies are only about 10,000 light-years across, so that they are separated by a distance 100 times their own size.

This does not mean that the galaxies are more or less *uniformly* distributed with the spacing of a million light-years. There does, in fact, exist considerable clustering, ranging from pairs of galaxies through clusters with fifteen or twenty members like the local group, up to clusters such as the one in Virgo containing several thousand galaxies.

In addition to studying the distribution of galaxies in space, Hubble classified them into various types, according to their general appearance. The main types are spiral, barred spiral, and elliptical, with a few per cent of irregular shape (Plate XI). The problem of understanding these different forms—whether they are related to age or some other property of a galaxy, and why the galaxies are distributed among these types in the observed proportions—is still unsolved.

Since 1936 many detailed studies of individual galaxies have been made, but on the whole the picture presented by Hubble still stands. The one major modification concerns the Cepheid variable method of determining distances. This method is based on the fact that there is a relation between the period of a Cepheid variable and its intrinsic brightness—the period-luminosity relation. Now Walter Baade, an American astronomer, discovered in 1952 that there are two types of Cepheids with different period-luminosity relations (Fig. 22). Distances within the Milky Way are unaffected by this discovery—they had been determined from Type II Cepheids, and so are correct. Unfortunately the Cepheids observed in other galaxies are of Type I, which means that their absolute brightness had been underestimated. In consequence their distances had also been underestimated. This increase in our knowledge of Cepheid variables has not lessened their value as distance indicators but has simply changed the value of the distances they indicate.

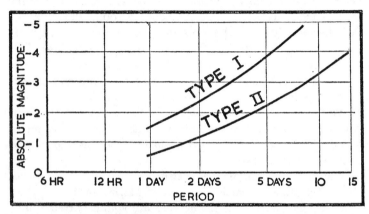

Fig. 22. The period-luminosity relation for the two types of Cepheid variables. The original relation (Fig. 12) referred to Type II Cepheids only.

The change has been a substantial one. Distances derived by Hubble should be multiplied by about 10, thus greatly increasing the scale of the universe. One consequence of this is that the Milky Way is no longer a giant among galaxies but is of average size, since the diameters of galaxies as deduced from their *angular* diameters must also be multiplied by 10. In this way we have lost the last vestiges of our privileged status. Copernicus dethroned the earth, Shapley the sun, and Baade the Milky Way. Since the local group of galaxies is a comparatively small cluster, the geocentric picture of the universe is completely discredited.

Further detailed improvements in our knowledge of galaxies

are bound to come, especially perhaps from the infant science of radio astronomy. But the real key to our understanding of the universe lies in a phenomenon which we have not yet mentioned—the phenomenon that first brought cosmology into the public eye—in other words, the *expansion* of the universe. But this deserves a chapter to itself.

CHAPTER V

THE EXPANDING UNIVERSE

The history of astronomy is a history of receding horizons.
E. P. HUBBLE

Introduction

The discovery that the universe is expanding took astronomers completely by surprise. They were quite unprepared for it, although we shall see later that they overlooked a broad hint. The most surprising feature was the great speeds which were involved, far greater than any they had previously encountered. The subsequent discovery that there is a pattern underlying these speeds was less surprising, since astronomers, like other scientists, are accustomed to finding patterns in nature.

The first observational hint of an expanding universe came from early measurements of the *radial* velocity of spiral galaxies—that is, their velocity *along* the line of sight. This motion should not be confused with the transverse velocity, or velocity at right angles to the line of sight. This latter velocity can be directly detected from the resulting change of position in the sky (proper motion), but how can radial velocity be detected? Our measurements of distance are far too inaccurate to reveal the resulting change in distance or angular diameter, except for objects in the solar system. Fortunately there exists an indirect method, which was devised in the middle of the last century. It is based on the so-called Doppler effect. This method is still our sole means of measuring the radial velocities of distant objects.

The Doppler Effect

This effect was discovered by the Austrian physicist Christian Doppler (1803–53) in 1842, in the course of his researches on acoustics. A familiar example of it occurs when a train passes by with its whistle blowing. Just as the train passes, the pitch of its whistle drops noticeably. Actually the whistle produces a higher note than normal continuously as the train approaches, and a lower note than normal as it recedes, but one notices only the sudden drop from high pitch to low.

This effect was explained by Doppler as follows (Fig. 23). The source of sound has a pitch that corresponds to emitting, let us say, 50 waves per second. If the source is stationary, then 50 waves reach the ear per second, so that the frequency of emission and of reception are the same. Suppose now that the source of sound is moving toward the ear. Then the last waves in a batch of 50 have less far to travel than the first wave. They will therefore arrive early. This means that more than 50 waves will be received in one second; in other words, the frequency of reception is greater than the frequency of emission. The pitch of the note has risen.

On the other hand, if the source of sound recedes from the ear, later waves have farther to travel, so the frequency of reception will be less than the frequency of emission. In this case the pitch of the note drops.

The essential feature of this effect is that the amount by which the frequency changes is related to the speed of the source. The faster the source recedes, for instance, the farther the later waves have to travel and the fewer arrive in one second. The drop in frequency is thus related to the speed of the source and is in fact proportional to it.

Doppler also suggested that light should behave in the same way as sound. In other words, if a source of light recedes from an observer the frequency of the light should be reduced.

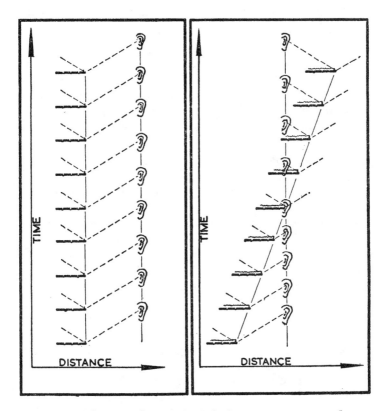

Fig. 23. The Doppler effect. (a) A stationary train whistles. The frequency (pitch) of the sound received is the same as the frequency of the sound emitted. (b) A moving train whistles. The frequency of reception is increased when the train approaches and decreased when it recedes.

Since the frequency of light determines its color, this drop in frequency leads to a color change; in fact the source is reddened. On the other hand, if a source of light approaches the observer the frequency of its light should be increased, so that the source looks bluer.

This idea led Doppler to attribute the various colors of the stars to their different radial motions. This explanation is incorrect. It overlooks the fact that for a given motion of the source the frequency change is much smaller for light than it is for sound. The reason for this is that the speed of light is very much greater than the speed of sound. The fact that later waves have farther to travel is then of less significance for light than it is for sound. In fact color changes are far too small to be detected unless the radial velocities of the stars approach the velocity of light.

Six years later the French physicist H. Fizeau suggested that, although the optical Doppler effect cannot be detected from color changes, it should be searched for in spectral lines, whose frequency can be measured very accurately by special techniques. Since a particular line arises from a specific type of atom, the frequency of a line in a stellar spectrum can be compared with its frequency when it arises from the same type of atom in the laboratory. The difference in frequency then enables the star's radial velocity to be inferred.

The first successful application of this idea followed twenty years later in 1868 when the English astronomer Sir William Huggins (1824–1910) measured the radial velocity of the star Sirius. He obtained the result that Sirius is receding from us at a speed of 29 miles per second. The corresponding shift in frequency is only one ten-thousandth of the actual frequency. The measurement of such a small shift presented at the time a very difficult problem. Huggins himself said of it: "It would be scarcely possible, even with greater space, to convey to the reader any true conception of the difficulties which present themselves in this work . . . and of the extreme care and caution which were needful to distinguish spurious instrumental shifts of a line from a true shift due to the star's motion." Thereafter Huggins studied the radial velocities of other bright stars and found that most of them were moving at about 30 miles per second.

The Velocities of Spiral Galaxies

A considerable time elapsed before these studies were extended to spiral galaxies. Indeed, the first successful observations were not made until 1912. In that year V. M. Slipher of the Lowell Observatory determined the shift in the spectral lines of the Andromeda galaxy and discovered that it is approaching us at a velocity of about 125 miles per second. When we recall that most stars move at about 30 miles per second, it will be seen that this is a remarkable result. Slipher went on to measure the spectra of other galaxies and found that most of them show a *red*-shift, which means that, unlike Andromeda, they are receding rather than approaching. The amount of the shift again implied large velocities. By 1914, Slipher had measured the spectra of thirteen galaxies, all but two of which are receding at around 100 or 200 miles per second.

These velocities were by far the largest that had ever been measured in astronomy. But worse was yet to come. By 1917 velocities of 400 miles per second had been registered, and even this record was soon surpassed. It is interesting to read a contemporary comment. Eddington wrote in 1923:

> One of the most perplexing problems of cosmogony is the great speed of the spiral nebulae. Their radial velocities average about 600 kilometres per second and there is a great preponderance of velocities of recession from the solar system. It is usually supposed that these are the most remote objects known (though this view is opposed by some authorities), so that here if anywhere we might look for effects due to general properties of the world.

Eddington then gave a list of the radial velocities of spiral galaxies as measured by Slipher up to February 1922 and continued:

The great preponderance of positive (receding) velocities is very striking; but the lack of observations of southern nebulae is unfortunate, and forbids a final conclusion. Even if these also show a preponderance of receding velocities the cosmogonical difficulty is not entirely removed. . . . It will be seen that two nebulae (including the great Andromeda nebula) are approaching with rather high velocity and these velocities happen to be exceptionally well determined.

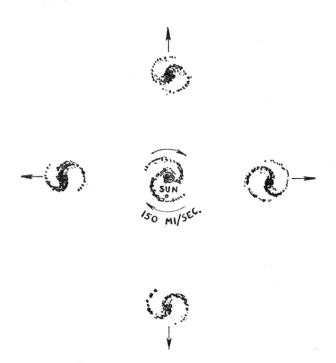

Fig. 24. *The radial velocities of galaxies relative to the sun and to the center of the Milky Way. The observed velocities of galaxies must be corrected for the sun's motion around the Milky Way, in order to obtain velocities relative to the Milky Way as a whole.*

THE EXPANDING UNIVERSE

Eddington's words remind us that at that time it had not yet been definitely established that the spiral galaxies lie outside the Milky Way. Hubble's discovery that they do dates from the next year, 1924. Further light on Slipher's velocities was shed by the discovery in 1926–27 that the Milky Way is in rotation. The sun's velocity around the center of the Milky Way is about 150 miles per second. Other objects in the Milky Way are also moving around its center, so that their radial velocity *relative* to the sun is much less than 150 miles per second (cf. Fig. 17). But objects *outside* the Milky Way do not share its rotation, so that their measured velocities must be corrected for the sun's motion if we want to know their velocity relative to the Milky Way as a whole (Fig. 24). When this was done the rapidly approaching galaxies which so worried Eddington were much slowed down. The irony of this is that after the correction the Andromeda galaxy had a velocity of

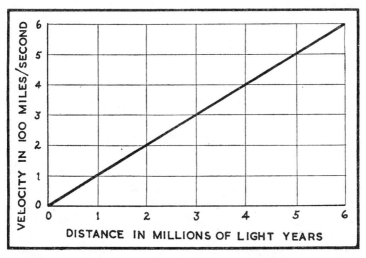

Fig. 25. *Hubble's first velocity-distance relation for galaxies. The velocity of recession is just proportional to the distance of a galaxy. Hubble's distances should be multiplied by 10.*

approach of only about 30 miles per second. Thus the first velocity measured by Slipher, which at the time seemed startlingly large, is actually no guide to the surprises that were to come.

The significance of Slipher's results was further clarified by Hubble's important discovery that the velocities of recession are by no means random. By using his measurements of the distances to spiral galaxies, Hubble established in 1929 that out to 6 million light-years the velocity of a galaxy is proportional to its distance (Fig. 25).

At first sight it might appear that the privileged status of the Milky Way had been restored by Hubble's discovery. However, it was quickly realized that Hubble's result does not imply that the Milky Way is a unique center of repulsion. On the contrary, a law of expansion in which velocity is just proportional to distance implies that any galaxy may be regarded as the center of expansion, and will observe the same law of recession (Fig. 26).

Hubble believed that the constant of proportionality in his law of recession was about 100 miles per second per million light-years. This means that for every million light-years of a galaxy's distance one must add another 100 miles per second to obtain its velocity of recession. This scale of velocities can be put in a more striking way, which serves to explain why Hubble's result created a great sensation. Let us trace the motion of the galaxies backward in time, assuming that each galaxy always had the velocity it has now. Then Hubble's result implied that 2 billion years ago all the galaxies were crowded on top of one another. This was striking not only in itself but because the ages of the earth and the sun were believed to be greater than 2 billion years!

Of course the assumption that the universe has been expanding at a constant rate may be false. In that case the crowding together may have occurred more than 2 billion years ago. This question could not be decided without a theory of the expansion. Meanwhile many people felt that the time of 2 billion years, which became known as Hubble's constant, had a basic significance for the universe as a whole.

This reaction may seem somewhat premature but later work tended to support it. By 1931, Hubble extended the range of validity of his law from 6 million to 150 million light-years. Finally, with the help of the American astronomer Milton L.

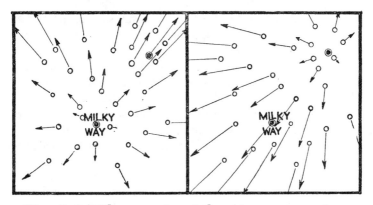

Fig. 26. (a) *The expansion of the universe as seen from different galaxies, according to Hubble's law. The recession velocity of a galaxy is proportional to its distance from the Milky Way.* (b) *The expansion of the universe as seen from another galaxy. The recession velocity is still proportional to the distance. Thus Hubble's law does not favor the Milky Way.*

Humason's new measurements of Doppler shifts, Hubble reached out to 240 million light-years, where the velocities of recession are about one seventh the velocity of light. This was the situation when Hubble published his book, *The Realm of the Nebulae,* in 1936.

Since then the 200-inch telescope has come into operation on Mount Palomar and improved techniques have been devised for detecting the light that is collected by the telescope. With their help W. A. Baum was able to measure in 1957 the Doppler shifts of galaxies estimated to be seven billion light-years away (on the new distance-scale), whose velocities of recession are two fifths the velocity of light. There is still no definite evidence of a deviation from Hubble's law, although a careful search was made to find one.[1] The only important change in Hubble's results came from the recognition of the large error in his distance scale. Hubble's constant is now believed to be more than 10 billion years. This is greater than the accepted ages of the earth and the sun so that there is no longer any difficulty in supposing that the universe was very dense 10 billion years ago.[2]

The great range of validity of Hubble's law naturally suggested to astronomers that they have observed a typical sample of the universe as a whole, and that nothing very different would emerge when they could see to greater distances. Moreover, clusters of galaxies appeared to be distributed more or less uniformly in space, which suggested that on this scale the universe has the same properties everywhere. Welcome support for this conclusion came from the theory of relativity, which showed that a uniform universe would in general obey Hubble's law.[3] So, with this theoretical backing, many astronomers concluded that they have observed a typical sample of a uniform universe; that the realm of the nebulae is indeed the universe itself.

The Future

What of the future? It appears that Baum's observations may represent the limit of what is possible with the 200-inch telescope, at least with existing techniques of light detection. Fortunately the exciting possibility has recently arisen that radio astronomy may be able to take over where optical as-

---

[1] A deviation was originally announced but, unfortunately, large selection effects make the results uncertain.

[2] But see Chapters XII–XV.

[3] Except at very large distances; hence the interest in deviations from Hubble's law.

tronomy leaves off. For galaxies have now been detected at 21 centimeters and also as radio sources emitting continuous radiation over a wide range of frequencies (Fig. 20). Some galaxies are relatively dull—such as our own and Andromeda. But there are also extragalactic sources which, by radio standards, are very bright—for instance, a pair of colliding galaxies (Plate XIII). It is probable that such colliding galaxies could be detected by radio telescopes even if they are too distant to be observed by the 200-inch telescope.

It appears, then, that we may be able to extend our observations of galaxies out to even greater distances than have yet been explored. I have suggested that when this happens there will be no new surprises. In the light of the numerous errors that have been described in these chapters, this may seem a very dangerous statement to make. Observers make it mainly because of the systematic nature of the red-shifts. But I think that the best reasons come from theory, which can explore depths of the universe beyond the reach of the most powerful telescopes. To this theory we now turn.

PART II

THE UNIVERSE IN THEORY

INTRODUCTION

In Part I of this book we saw that many important discoveries about distant regions of the universe have been made by the methods of observational astronomy. The great success of these methods represents a technical triumph of the first order. But it is not a triumph that need take us by surprise—we expect to obtain information about distant objects when we look through very powerful telescopes. What *is* surprising, perhaps, is that we can also obtain information about these objects without using telescopes at all—simply by reasoning about phenomena that occur in the laboratory.

What makes this possible is the fact that the universe is not a collection of independent objects. Its different regions strongly influence one another, as in a machine all of whose parts are closely linked together. If we know the way in which such a machine works we can often tell, for instance, whether distant cogwheels are rotating, simply by inspecting the position of a nearby lever. In the same way observations of our own immediate neighborhood can be used to obtain information about distant regions of the universe. The regions concerned are more distant than those which have hitherto been explored by telescopes. In consequence, if we know how the links operate we can extend our knowledge of the universe to regions beyond the reach of existing telescopes.

At the same time this procedure gives insight into local phenomena by revealing the strong influence that distant matter

INTRODUCTION

exerts on nearby matter. This situation presents us with three interdependent problems: discovering how the links operate, determining the properties of distant regions of the universe, and improving our understanding of local phenomena.

This triple problem is studied in the next few chapters. We then go on to discuss the problems which arise from the *uniqueness* of the universe. The universe is obviously unique, since by definition it consists of the totality of things. This nevertheless raises problems, and for the following reason. Scientists normally have available for study many instances of any particular phenomenon. By comparing these instances with one another they are able to distinguish between the fundamental and the accidental aspects of the phenomenon. For example, by comparing many instances of motion under gravitation, Newton discovered that the shape of an orbit is fundamental but its size is not.

Now with only one universe available for study, we have no basis for distinguishing between its fundamental and its accidental features. Two choices are then open to us. We can regard all its features as either equally fundamental or equally accidental. For my part, I believe that the aim of science should be to show that no feature of the universe is accidental. This may seem unexceptional, but nevertheless it has very definite consequences for the remote past and future of the expanding universe. For, according to modern theory, only if the universe as a whole is unchanging in time, the continual creation of new matter compensating for the expansion, will its main properties be fundamental. Whether the universe actually has this character may be decided by observation in the next few years.

CHAPTER VI

OLBERS' PARADOX

> or if they list to try
> Conjecture, he his Fabric of the Heav'ns
> Hath left to thir disputes, perhaps to move
> His laughter at thir quaint opinions wide
> Hereafter, when they come to model Heav'n
> And calculate the Stars.
> MILTON, *Paradise Lost*

Introduction

The first discovery of a link connecting us to distant regions of the universe also marks the birth of scientific cosmology. It occurred in 1826 with the publication of a brilliant paper by the German astronomer Heinrich Olbers (1758–1840), in which he propounded a remarkable paradox. But this contribution to cosmology was an isolated one. Not until the end of the century were further calculations made involving the universe as a whole, and by that time Olbers' paradox had been largely forgotten. This was unfortunate, for had it been remembered the discovery that the universe is expanding would not have taken people by surprise—indeed, they would have expected it. As it was, the expansion was discovered by direct observation and only afterward seen to resolve Olbers' paradox.

The Paradox

Olbers' starting point was an attempt to calculate the total amount of light reaching us from all the stars. Since it is dark at night it was clear to him that this total is only a small fraction of the light coming from the sun. He set out to calculate this fraction in terms of the properties of the stars—their brightness, number, and distance from us. This is a very natural problem for an astronomer to tackle—the result would be of interest in itself, and of use in other problems—but, nevertheless, it must have seemed to Olbers at the time to be a purely routine calculation. Quite a surprise was in store for him.

Olbers' calculation is easy to follow if it is broken down into a number of steps. In the first step we consider the contribution of one star, which, of course, involves the inverse square law of apparent brightness.[1] We shall find it convenient to look at this law in terms of the following picture. The rays of light emitted by the star can be regarded as made up of small particles (nowadays known as photons) which move in straight lines. Each photon carries a certain amount of energy, and since energy is conserved, a photon has the same energy all the way along its track. For simplicity, we shall assume that this energy is the same for all photons, although in practice many different energies are represented.

Now let us compute the total amount of energy that reaches the eye in one second—this just corresponds to the apparent brightness of the star. This apparent brightness is the product of three factors—the energy carried by each photon, the number of photons arriving along each track in one second, and the total number of tracks that reach the eye. This last number is proportional to the *solid angle* which the tracks form at the eye (Fig. 27).

[1] Strictly speaking, a correction must be made for the shape of the star unless it is effectively a point source.

OLBERS' PARADOX

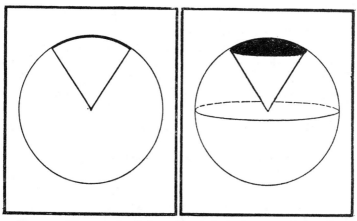

Fig. 27. Plane and solid angles. (a) A plane angle can be measured by the length it cuts off from the circumference of a circle of unit radius. (b) A solid angle can be measured by the area it cuts off from the surface of a sphere of unit radius.

We can now compare the apparent brightnesses of similar stars which are at different distances but which are not in relative motion (Fig. 28). The energy carried by each photon

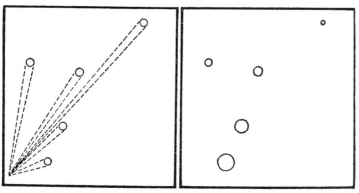

Fig. 28. The apparent brightnesses of intrinsically similar stars are proportional to the solid angles they subtend at the eye.

is the same for all the stars, if they are similar; and so is the number of photons arriving per second along each track. What has changed is the third factor, the solid angle formed by the tracks. This solid angle is less for distant stars than for nearby ones, so that distant stars look fainter. The dependence of the solid angle on the distance of the star thus determines the way in which the apparent brightness of a star depends on its distance. This dependence, in fact, corresponds to the inverse

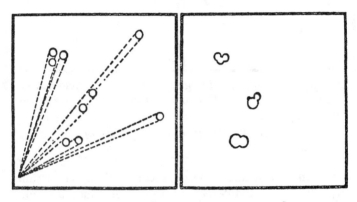

Fig. 29. The total brightness of a set of similar stars is proportional to the total solid angle they subtend at the eye.

square law,[2] but it is best to think directly in terms of solid angles. To sum up: The *intrinsic* brightness of a star determines the energy of each photon and the rate of emission (and arrival) of the photons. The *apparent* brightness of a star of given intrinsic brightness is then determined entirely by the solid angle which the star subtends at the eye—indeed, it is proportional to this solid angle.

We now come to the second step in Olbers' calculation. Consider what happens when several stars are simultaneously present (Fig. 29). If these stars have the same *intrinsic* bright-

[2] Apart from the shape factor mentioned in the previous footnote.

ness, then the total amount of energy reaching the eye per second is just proportional to the total solid angle which these stars subtend at the eye.

An essentially new situation arises in the third step. Here one star is so close to the eye that it completely fills its field of vision. The solid angle subtended by the star is now effectively the same whatever the distance of the star, *so long as it continues to fill the field of vision*. This means that its apparent brightness is independent of its distance, and indeed is the

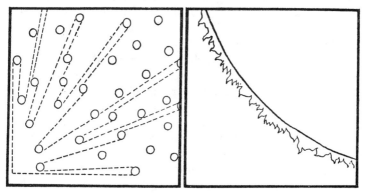

Fig. 30. (a) A set of similar stars which fill the field of vision. (b) One star which fills the field of vision.

same as if the eye were right at *the surface of the star*. All the eye can see in such a case is a uniform blur of light filling its whole field of vision, and it has no means of telling how far away the source of light is.

The same situation holds for a number of stars of equal intrinsic brightness, if in combination they fill the field of vision (Fig. 30). In this case, too, the eye will see the same brightness as at the surface of any one of the stars. To put it more graphically, if each line of sight eventually ends in a star, then no matter how long these lines of sight may be the eye will not be able to distinguish the different stars but will think itself to be right at the surface of one of them.

Equipped with these results, we can now consider the last step in Olbers' calculation—the determination of the total amount of light at the surface of the earth arising from all the stars except the sun. Of course stars vary considerably in their intrinsic brightness, but we will not go far wrong if we replace each star by one of average brightness. The more important question is: How many stars are there? Olbers naturally assumed, to begin with, that the stars are distributed more or less uniformly throughout space. But in that case every line of sight will eventually meet a star. This means that *the amount of light near the earth should be just the same as at the surface of an average star.* The temperature of space would then be about 6000 degrees![3]

This is the sort of result that a scientist is delighted to obtain. A straightforward calculation, based on a few simple and natural assumptions, leads to a result which is in violent disagreement with observation. This means that at least one of the assumptions must be quite wrong. This in itself is an important discovery, but, more important still, there is the prospect that the wrong assumption can be identified and corrected. Let us try to do this for Olbers' paradox.

Resolution of the Paradox

Olbers made the following assumptions:

(i) Every line of sight ends in a star of average intrinsic brightness.

(ii) There is no interstellar fog absorbing some of the light.

(iii) There are no systematic motions of the stars.

Which of these simple assumptions should be altered? It was at this point that Olbers missed his great chance, for he chose to alter the second assumption and to explain the dark-

---

[3] This conclusion can be reached immediately by thermodynamical considerations, since in the absence of a sink of radiation the universe would settle down into a state of equilibrium, in which radiation has the same temperature everywhere.

ness of the night sky in terms of an interstellar fog. Although we now know that such a fog does exist, *by itself* this explanation is inadequate. For in the course of time the fog would have been heated up by the light it absorbed until it became just as hot as the surface of a star. It would then radiate as much light as it absorbed, and so would not serve the required purpose. Had Olbers realized this, he would have been forced to alter one of his other assumptions.

In order to decide which assumption is false it is very helpful to know which stars are the main contributors to the total light calculated by Olbers. Now the average distance between the stars, 10 light-years, is far greater than the diameter of a star, 3 light-seconds. Consequently most lines of sight stretch a very long way before they meet a star. In fact, if the ratio of

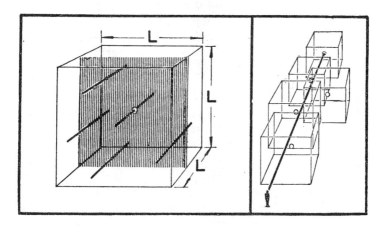

Fig. 31. (*a*) *Calculation of the average length of a line of sight. Construct a box round each star of side equal to* L, *the average spacing between the stars. The chance,* S^2, *that a line of length* L *cuts a star is equal to the area of a star divided by* L^2. (*b*) $1/S^2$ *lines of length* L *placed end to end are fairly certain to cut a star. Thus the average length of a line of sight is about* L/S^2, *which is* 10^{17} *light-years.*

separation to size is S, the average length of a line of sight is S^2 times the separation (Fig. 31). This means that the average length of a line of sight is 10^{17} light-years! In other words, *most of the responsible stars are a tremendous distance away from us.*

Here we meet for the first time an example of the powerful influence exerted on us by distant regions of the universe. Of course our immediate task is to reduce this influence, in the sense that we must discover something about distant stars which drastically decreases their contribution to the light of the night sky. But when we have succeeded, the basic fact will remain: it is dark at night only because distant stars are behaving in a special way. This is an example of the three interdependent problems mentioned in the Introduction: we have found that there is an optical link connecting different regions of the universe and that it can be used both to obtain information about distant stars and to understand a local phenomenon, in this case the alternation of night and day.

Now let us try to obtain this information about distant stars. Which of Olbers' assumptions should be altered? Consider, first, assumption (i), that every line of sight ends in a star of average intrinsic brightness. We might assume instead that the intrinsic brightness of the stars decreases with distance, so that 10^{17} light-years away the stars are extremely cold and dark. Since these stars are being observed as they were 10^{17} years ago, this would mean that they were very faint in the distant past, or else were formed relatively recently. If they ultimately become as bright as nearby stars the escape from Olbers' paradox would be only temporary, and the universe would be destined to reach an uncomfortably high temperature. This distasteful conclusion could be avoided by supposing that distant stars never become very hot or indeed do not exist at all. Unfortunately, this would mean that the universe is grossly non-uniform in structure. It would be more satisfying if we could resolve Olbers' paradox without having to attribute completely different properties to different regions of the universe.

One way of doing this was discovered in 1922 by the Swedish statistician C. V. L. Charlier. He suggested that the structure of the universe might be infinitely complex, in the following sense. Just as stars are grouped together into galaxies, and galaxies are grouped into clusters, so these clusters may

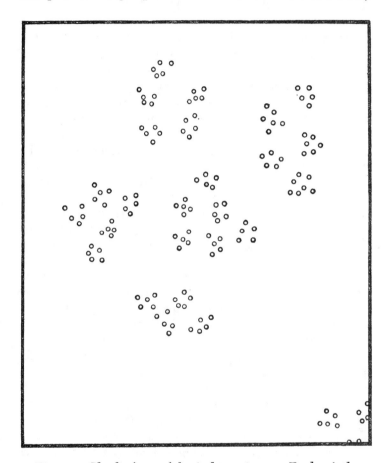

Fig. 32. Charlier's model of the universe. Each of the small circles is a galaxy, which belongs to a cluster. These clusters are themselves clustered, and so on indefinitely. This model will explain Olbers' paradox if the average density of matter in a cluster is less for higher-order clusters than for lower-order ones.

themselves be grouped into superclusters, these superclusters into super-superclusters, and so on indefinitely (Fig. 32). If the distance between high-order superclusters is made sufficiently large the calculated amount of light in the night sky can be reconciled with observation.

This model of the universe is in a sense uniform but it is very complicated. Fortunately a less complex resolution of Olbers' paradox can be found which still implies that the universe has a uniform structure. This involves the rejection of assumption (iii), that there are no systematic motions of the stars. For in a uniform universe we can reduce the light from distant stars by assuming that *they are receding in a uniform way*. Since such a uniform motion of recession is actually evident in the Doppler shifts of galaxies, Charlier's supercluster model is no longer needed to resolve Olbers' paradox. In fact, it is precisely this Doppler shift which reduces the amount of light. This is obvious in our photon picture, since Doppler's explanation of the decrease in the number of waves arriving per second implies that there is also a decrease in the number of photons arriving per second (Fig. 33). In addition, although this is not so obvious, the energy of each photon is also decreased. For both these reasons the contribution of distant stars is reduced.[4] Now, in a uniform expansion a source recedes at a rate proportional to its distance (Fig. 26), so the greater its distance the more its contribution is reduced. The main effect of the expansion of the universe is thus to suppress the light from the very distant stars that led to Olbers' paradox.

We conclude that *it is dark at night because the universe is expanding*. Moreover, it must be expanding at just the rate required to account for the observed intensity of the night sky. In practice, the expansion is so fast that most of this intensity is of local-origin-scattered sunlight and radiation from the stars in our own galaxy. These local sources can be allowed for approximately, and the actual rate of expansion—that is, Hubble's constant—thereby inferred. Olbers would not in his time have

---

[4] It is still further reduced by aberration.

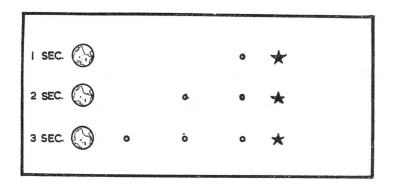

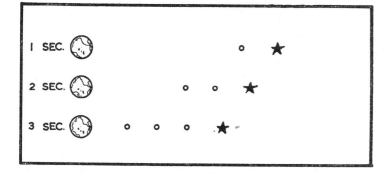

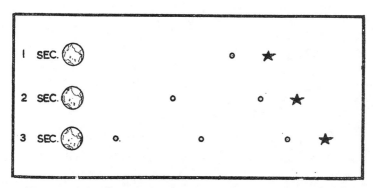

Fig. 33. (a) Photons from a stationary star. (b) Photons from an approaching star arrive at a greater rate. (c) Photons from a receding star arrive at a smaller rate.

been able to distinguish between local sources and receding galaxies, but he might still have predicted the expansion of the universe, and even made a very rough estimate of Hubble's constant, a hundred years ahead of the observers. His failure to do so is one of the greatest missed opportunities in the whole history of science.

CHAPTER VII

MACH'S PRINCIPLE

> That there are distinct orders of Angels, assuredly I believe; but what they are, I cannot tell; Dicant qui possunt; si tamen probare possunt quod dicunt, saies that Father, Let them tell you that can, so they be able to prove, that they tell you true. They are Creatures, that have not so much of a Body as flesh is, as froth is, as a vapor is, as a sigh is, and yet with a touch they shall molder a rock into lesse Atomes, then the sand that it stands upon; and a milstone into smaller flower, then it grinds.
>
> JOHN DONNE

Introduction

Olbers' paradox shows that there is an optical link connecting us to distant regions of the universe. The idea underlying Mach's principle is that we are also joined to distant regions by a *mechanical* link, which would mean that the universe as a whole exerts a strong influence on the *motion* of nearby matter. According to Mach, this influence reveals itself in the inertial properties of matter. If this is true it is of the utmost importance, since the inertia of matter is probably its most fundamental property.

Mach's principle has given rise to considerable controversy, some of it rather outspoken. My own view is that Mach is right, for reasons which I shall develop in this chapter. The next two chapters will then show how Mach's principle can be used, like Olbers' paradox, to determine some important properties of distant regions of the universe.

The Inertia of Matter

We owe our present definition of the inertia of matter to Galileo. He seems to have been the first person to realize that it is not the velocity of a body but its *acceleration* which signifies that there are forces acting on it. An undisturbed body remains at rest or moves with constant velocity. In practice, "undisturbed" moving bodies are slowed down by friction, but when this friction is drastically reduced, as on an ice rink, their velocity is maintained for a long time. This makes Galileo's discovery seem plausible.

The fact that we have to exert a force on a body in order to accelerate or decelerate it is usually expressed by saying that the body has inertia. This nomenclature corresponds to our picture that matter "resists" attempts to change its velocity, and that this resistance must be overcome by the exertion of a force. The notion of inertia can very easily be made quantitative: we can say that the *amount* of inertia possessed by a body is measured by the force required to change its velocity by a given amount. For instance, it obviously requires a greater force to stop a motorcar than it does to stop a bicycle moving at the same speed. A motorcar thus has more inertia than a bicycle—how much more is measured by the ratio of the two forces involved.

Once we can measure the inertia of matter we are in a position to ask some new questions. The most important, perhaps, is the following: Is the inertia of a given body always the same, or does it change when other bodies are brought near it? In fact no such change has ever been detected experimentally. This negative result led Newton to believe that inertia is an *intrinsic* property of matter, completely independent of its environment.

Despite the authority with which Newton's views came to be invested, another school of thought arose, which is mainly associated with the name of the Viennese physicist and philosopher Ernst Mach (1838–1916). According to Mach, a body has inertia because it interacts in some way with all the matter in the universe. Nearby matter would then make a very small contribution to the total amount of inertia, rather than no contribution at all. Mach's view exerted a great influence on Einstein, who gave it the name "Mach's principle." Before discussing this principle we shall give a detailed account of Newton's views on inertia.

Newton's Theory of Inertia

NEWTON'S SECOND LAW OF MOTION. Newton's theory of inertia is based on his famous second law of motion. This law states that the force acting on a body is proportional to its acceleration (and is in the same direction). The constant of proportionality is called the inertia or inertial mass of the body, so that Newton's second law can be expressed in the familiar form: force is mass times acceleration.

When we look carefully at this law we find a curious difficulty. For, while the force acting on a body is objectively determined by whatever is exerting the force, the value of the acceleration depends on how it is measured, that is, on which body is taken as providing a standard of rest. Consider, for instance, the behavior of the earth under the action of the sun's gravitational force. In Fig. 34a we have taken the sun to be at rest and the earth to be moving around it once every year. In such a circular motion it is known that the acceleration is always in the direction of the center—that is, toward the sun. This acceleration implies that the earth's velocity is swinging around, so that it keeps moving along its circular path. The earth's acceleration is thus in the same direction as the sun's gravitational force, in accordance with Newton's second law. Moreover, the amount of acceleration is proportional to the strength of this force, so that Newton's second law is completely satisfied.

On the other other hand, suppose we take the earth to be at

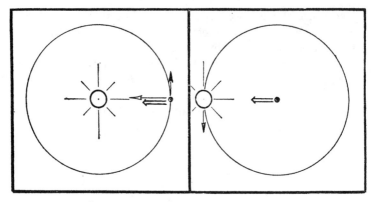

Fig. 34. The sun's gravitational force on the earth and Newton's second law of motion. (a) The circular motion of the earth relative to the sun. The earth's velocity (arrow with black head) is tangential to the circle. Its acceleration (arrow with double stem) is toward the sun, as is the sun's gravitational force (arrow with white head). Force and acceleration are thus in the same direction, in accordance with Newton's second law. (b) The motion of the sun relative to the earth. In this case the sun's gravitational force produces no acceleration at all.

rest and the sun to be moving around it in a circle (Fig. 34b).[1] In this case Newton's second law is not satisfied, since the sun's gravitational force is producing no acceleration of the earth at all!

A similar example of this difficulty is provided by the motion of artificial satellites. The ones which have been launched so far have circled the earth in an hour or two. But the farther

---

[1] This appears to ignore Copernicus, but we are deliberately going back to first principles. Moreover, the importance of Copernicus lay in his rejection of the geocentric view of the universe, rather than in his suggestion that the sun does not move. This suggestion has been superseded both by the development of astronomy (see Part I) and by the theories of inertia described in this and the following chapters.

MACH'S PRINCIPLE

out a satellite is, the longer it takes to complete its orbit. At a certain height it will take just twenty-four hours. If a satellite at this height were to move parallel to the equator and in the same direction as the earth rotates, it would always be above the same point of the earth's surface (Fig. 35). Someone looking up would see a body *at rest* above his head, hovering with no visible means of support!

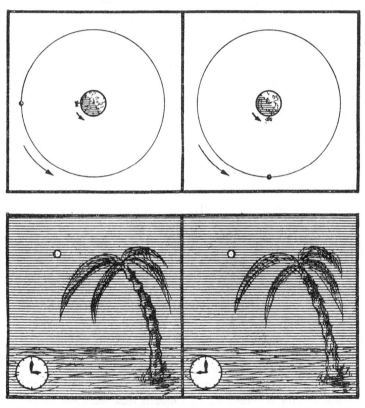

Fig. 35. (a) *A satellite rotating parallel to the equator with a period of twenty-four hours.* (b) *At the equator the satellite is seen to be constantly hovering overhead with no visible means of support.*

These examples show that Newton's second law applies only if the accelerations of bodies are measured in a special way. Since Newton believed his law to be fundamental, he supposed that accelerations measured in such a way that his law applies are of particular significance, and he called them *absolute*. Newton's second law should now be amended to read: force is mass times absolute acceleration. Those bodies on which no forces act will then have no absolute acceleration. Such bodies are said to constitute an *inertial frame of reference* or simply an inertial frame, because accelerations measured relative to them will be absolute accelerations. Consequently for Newton's second law to be satisfied accelerations must be measured relative to an inertial frame of reference. An example of such a frame is shown in Fig. 34a, which depicts the earth as revolving around the sun once a year; in this case Newton's second law is satisfied.

INERTIAL FORCES. Inertial frames naturally play a fundamental role in Newton's theory. Nevertheless, he often found it convenient to use a *non*-inertial frame of reference—that is, to measure accelerations relative to some body whose absolute acceleration is not zero. Examples of such non-inertial frames are shown in Figs. 34b and 35b. This procedure leads, of course, to anomalies, in particular that a force may produce no acceleration at all. Nevertheless, Newton was able to adapt his law of motion to fit this situation by postulating the existence of some *additional* forces, which do not have a physical origin in material objects. These additional forces, commonly called inertial forces, are needed to compensate for measuring accelerations relative to a non-inertial frame of reference.

To see how this device works, consider again Fig. 34b. In this figure the earth has no acceleration despite the action on it of the sun's gravitation. This means that we must be using a non-inertial frame of reference. By comparing this frame with the inertial frame of Fig. 34a we see that it is non-inertial because it is absolutely rotating. If we want to be able to apply Newton's second law in this absolutely rotating frame of reference we must introduce an inertial force acting on the earth which just cancels the gravitational force of the sun (Fig. 36a).

A similar force must be introduced in Fig. 35b, which shows a stationary satellite hovering overhead. The earth must be in a state of absolute rotation, so that a frame of reference at

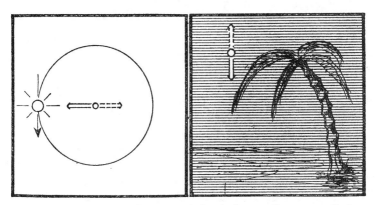

Fig. 36. Centrifugal forces. (a) If the earth is regarded as at rest, an inertial (centrifugal) force must be acting on it to balance the gravitational force of the sun. (b) A similar force is needed by the stationary satellite to balance the gravitational force of the earth.

rest relative to the earth will also be absolutely rotating. An inertial force acting on the satellite must then be introduced to counteract the earth's gravitation (Fig. 36b).

The inertial force which we have here introduced to compensate for using an absolutely rotating frame of reference is known as a centrifugal force, because it acts in a direction away from the center of rotation. This force is well known to every driver who has taken a corner too fast. What is less well known is that it is not the only inertial force which has to be introduced in a rotating frame of reference. If a body is *moving* in such a frame of reference an additional inertial force will act on it.

To see this, let us go back to our original inertial frame of reference, in which the earth revolves around the sun once every year (Fig. 34a). Now suppose that two small balls are shot out from the earth, one in the direction of the earth's motion and one in the opposite direction (Fig. 37a). The ball which is shot out behind the earth moves more slowly around

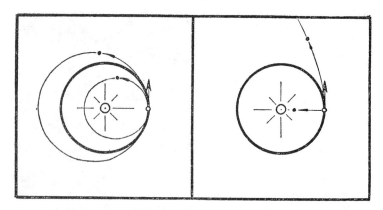

Fig. 37. (a) *The orbits of two balls thrown out from the earth (whose own orbit is the thick black circle). The ball in the outer orbit was thrown forward (that is, in the direction of the earth's motion), and the one in the inner orbit was thrown backward.* (b) *The extreme case in which the ball thrown backward falls directly into the sun, while the one thrown forward escapes from it.*

the sun than the earth itself does (we are here neglecting the earth's gravity). Consequently it cannot move in the same orbit as the earth, since its lower velocity would then imply that its acceleration toward the sun is less than the earth's. The sun's gravity therefore pulls it radially inward, and it moves in an orbit inside the earth's. An extreme example of this occurs when the ball is thrown out at a speed equal to that of the earth—that is, about 20 miles per second (Fig. 37b). For the ball will then be momentarily at rest relative to the sun and will simply fall directly toward it. A less extreme example of this type of inward motion comes from the behavior of artificial earth satellites, which are slowed down by air resistance. As a result they, too, have a circular motion which is too slow, so that they are pulled inward toward the earth—eventually to be burned up in the dense regions of the atmosphere.

What happens to the other ball, the one which is shot out ahead of the earth? It cannot remain in the earth's orbit either, but this time because it is moving too *fast*. The sun's attraction is now too small to keep it in the earth's orbit, so that it moves in an orbit outside the earth's (Fig. 37a).[2] If it is thrown forward at 20 miles per second, as a partner to the one that fell directly into the sun, it will escape from the sun altogether (Fig. 37b).

What do these motions look like in the non-inertial rotating frame of Fig. 34b? In this frame the earth is at rest and the

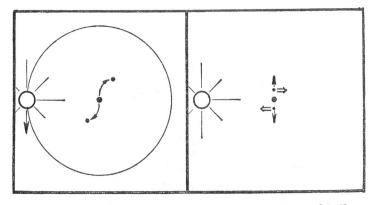

Fig. 38. Coriolis force. (a) The motion of the ejected balls relative to the earth. (b) The inertial (Coriolis) force that must be introduced to account for the motion of the balls. It acts in opposite directions on balls with oppositely directed velocities.

[2] Since these words were written, the Russian Lunick has been launched into just such an orbit.

two balls will leave it as in Fig. 38a, which also shows part of their subsequent motion. If Newton's second law is to apply to these motions we must introduce an additional inertial force which acts only on *moving* bodies like the balls, and not at all on the earth. Moreover, it must act in *opposite* directions on bodies with oppositely directed velocities (Fig. 38b). This somewhat complicated inertial force is known as the Coriolis force.

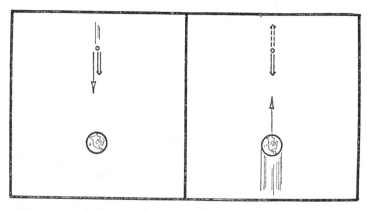

Fig. 39. (a) A body falling toward the earth. The earth's gravitational force (arrow with double stem) produces an acceleration (arrow with single stem) in accordance with Newton's second law. (b) The same situation from the body's point of view. Here the body is at rest and the earth is accelerating. An inertial force must now act on the body to counteract the gravitational force of the earth.

As a final example of inertial forces, consider an accelerated body which falls in a straight line toward the earth (Fig. 39a). If we take as our frame of reference the non-inertial frame in which the body is permanently at rest (Fig. 39b), we must introduce an inertial force as shown. (This type of inertial force, related to acceleration in a straight line, is well known to every driver who has braked too fast.) How large must this inertial force be? To answer this, we recall that the earth's force is equal to the mass of the body times its *absolute* acceleration. It follows that the inertial force which we have to introduce must also be equal to the mass of the body times its absolute acceleration, if it is to cancel the gravitational force of the earth. For a similar reason centrifugal force is also proportional to the mass of the body it acts on. This property of inertial forces will be of fundamental importance later on.

ABSOLUTE SPACE AND ABSOLUTE MOTION. The upshot of our discussion so far is that all frames of reference are not equivalent for the formulation of laws of motion. There exists in nature a *special* set of reference frames, namely those in which Newton's second law holds good without inertial forces having to be introduced. The existence of such special (inertial) frames raises two related questions. First, why should there be a special set of frames in nature, and, second, what factors decide *which* frames are the special ones? Newton answered these questions by postulating the existence of an entity which he called "absolute space." Absolute motion is then defined as motion relative to absolute space, and the acceleration referred to in his second law is acceleration relative to absolute space.

The problem that now faced Newton was that absolute motion is not *directly* observable. How can it be identified among the welter of relative motions that are actually observed? Newton asserted that this is "a matter of great difficulty," but he added, "the thing is not altogether desperate." In his *Principia* (1686) he described a simple experiment for detecting, in particular, absolute rotation. This experiment has played a leading role in all subsequent discussions of the subject and is of such fundamental and historical importance that I shall quote his account of it in full.

> The effects which distinguish absolute from relative motion are, the forces of receding from the axis of circular motion. For there are no such forces in a circular motion purely relative, but in a true and absolute circular

motion, they are greater or less, according to the quantity of the motion. If a vessel, hung by a long cord, is so often turned about that the cord is strongly twisted, then filled with water, and held at rest together with the water; thereupon, by the sudden action of another force, it is whirled about the contrary way, and while the cord is untwisting itself, the vessel continues for some time in this motion; the surface of the water will at first be plain, as before the vessel began to move; but after that, the vessel, by gradually communicating its motion to the water, will make it begin sensibly to revolve, and recede by little and little from the middle, and ascend to the sides of the vessel, forming itself into a concave figure (as I have experienced), and the swifter the motion becomes, the higher will the water rise, till at last, performing its revolutions in the same times with the vessel, it becomes relatively at rest in it. This ascent of the water shows its endeavour to recede from the axis of its motion; and the true and absolute circular motion of the water, which is here directly contrary to the relative, becomes known, and may be measured by this endeavour. At first, when the relative motion of the water in the vessel was greatest, it produced no endeavour to recede from the axis; the water showed no tendency to the circumference, nor any ascent towards the sides of the vessel, but remained of a plain surface, and therefore its true circular motion had not yet begun. But afterwards, when the relative motion of the water had decreased, the ascent thereof towards the sides of the vessel proved its endeavour to recede from the axis; and this endeavour showed the real circular motion of the water continually increasing, till it had acquired its greatest quantity, when the water rested relatively in the vessel. And therefore this endeavour does not depend upon any translation of the water in respect of the ambient bodies, nor can true circular motion be defined by such translation. There is only one real circular motion of any one revolving body, corresponding to only one power of endeavouring to recede from its axis of motion, as its proper and adequate effect; but relative motions, in one and the same body, are innumerable, according to the various relations it bears to external bodies, and, like other relations, are altogether destitute of any real effect, any otherwise than they may perhaps partake of that one only true motion.

We can sum up Newton's interpretation of his experiment by saying that absolute rotation has nothing to do with the relative rotations which are directly observed, and that, nevertheless, we *can* determine experimentally the amount of absolute rotation possessed by a body. All we have to do is to measure the curvature of a water surface rotating with the body.

This method of detecting the absolute rotation of a body consists essentially of measuring the amount of centrifugal force acting on it. We can also devise methods based on a measurement of the Coriolis force. The best-known example of such methods is the Foucault pendulum, which was first used amid great acclaim by the French physicist Léon Foucault at the Paris Exhibition of 1851, to demonstrate the absolute rotation of the earth. The Foucault pendulum, which is now a familiar sight in science museums, consists of a weight suspended so as to be free to swing in any direction (Plate XIV). For simplicity, let us consider such a pendulum swinging at one of the poles. At other latitudes it will have a more complicated motion, but the principle is the same. Since the pendulum is swinging from a universal joint, the plane of its motion will remain fixed in absolute space, while the earth rotates underneath (Fig. 40a). To an observer on the earth, the pendulum will appear to swing around once in twenty-four hours (Fig. 40b). This motion of the pendulum can then be attributed to the action on it of a Coriolis force (Fig. 40c). Thus the rate of rotation of the earth can be measured by observations confined to its surface—no recourse is needed to some other body relative to which the rotation must be measured.

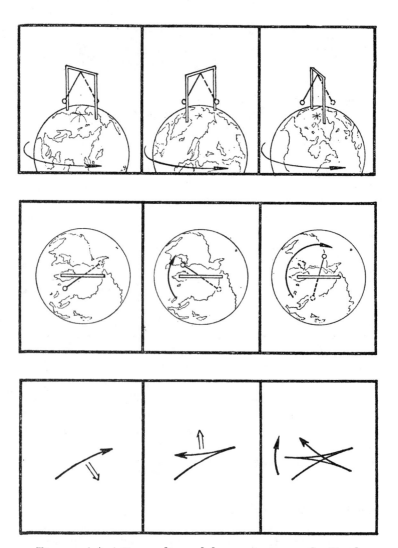

Fig. 40. (a) A Foucault pendulum swinging at the North Pole. It moves in a fixed plane while the earth rotates underneath. (b) To an observer on the earth, the plane of the pendulum appears to swing around as shown. (c) The apparent motion of the plane of the pendulum is due to the Coriolis force (white arrow) whose transverse action swings the pendulum around.

MACH'S PRINCIPLE

Berkeley's Criticism of Newton's Theory

Those among Newton's contemporaries who did not believe in absolute space still had to accept the result of his experiment with the bucket of water—they certainly could not adopt the well-known device of disappointed theorists and claim that it had been performed inaccurately! Instead, the experiment was ingeniously reinterpreted. The leading figure behind this move was the great Irish philosopher Bishop Berkeley. His position can be summed up by the following statement: all space corrupts and absolute space corrupts absolutely.

Berkeley objected to the idea of absolute space on the ground that it is unobservable. Moreover, he claimed that in interpreting the bucket experiment Newton had gone beyond the facts. As Berkeley emphasized, the relative motion of the water and the bucket is not more important than the relative motions of the water and other objects. Indeed, the only relevance of the bucket is as a vessel for holding the water. What one should deduce from the experiment is that centrifugal forces appear only when a body rotates *relative to the stars*. This is to be expected, since it is precisely the presence of such other matter which gives meaning to the concept of motion. As Berkeley put it, writing twenty years after the publication of the *Principia*:

> If every place is relative then every motion is relative and as motion cannot be understood without a determination of its direction which in its turn cannot be understood except in relation to our or some other body. Up, down, right, left, all directions and places are based on some relation and it is necessary to suppose another body distinct from the moving one . . . so that motion is relative in its nature, it cannot be understood until the bodies are given in relation to which it exists, or generally there cannot be any relation, if there are no terms to be related.
>
> Therefore if we suppose that everything is annihilated

except one globe, it would be impossible to imagine any movement of that globe.

Let us imagine two globes and that besides them nothing else material exists, then the motion in a circle of these two globes round their common centre cannot be imagined. But suppose that the heaven of fixed stars was suddenly created and we shall be in a position to imagine the motion of the globes by their relative position to the different parts of the heaven.

In writing this, Berkeley showed himself to be many years ahead of his time. His great contemporary, the Swiss mathematician Euler (1707–83), for instance, considered that the alleged influence of the fixed stars was "very strange and contrary to the dogmas of metaphysics." This view was echoed by many writers, but nothing of any significance was added to the discussion until the advent of Mach, a hundred and fifty years later.

Mach's Approach

Mach's approach to the problem of inertia was only a slight elaboration of Berkeley's, and it is important largely because it stimulated a rediscussion of the problem at a time when Newton's authority was unquestioned. Mach's criticisms of Newton's laws of motion are more detailed than Berkeley's, but as regards centrifugal force his standpoint is the same. In 1872, Mach wrote:

For me only relative motions exist. . . . When a body rotates relatively to the fixed stars, centrifugal forces are produced; when it rotates relatively to some different body and not relative to the fixed stars, no centrifugal forces are produced. I have no objection to just calling the first rotation so long as it be remembered that nothing is meant except relative rotation with respect to the fixed stars.

In one point he is more explicit than Berkeley, when he says:

> Obviously it does not matter if we think of the earth as turning round on its axis, or at rest while the fixed stars revolve round it. Geometrically these are exactly the same case of a relative rotation of the earth and the fixed stars with respect to one another. But if we think of the earth at rest and the fixed stars revolving round it, there is no flattening of the earth, no Foucault's experiment, and so on—at least according to our usual conception of the law of inertia. Now one can solve the difficulty in two ways. Either all motion is absolute, or our law of inertia is wrongly expressed. I prefer the second way. The law of inertia must be so conceived that exactly the same thing results from the second supposition as from the first. By this it will be evident that in its expression, regard must be paid to the masses of the universe.

According to Mach, then, inertial frames are those which are unaccelerated relative to the "fixed stars"—that is, relative to some suitably defined average of all the matter in the universe. Moreover, matter has inertia only because there is other matter in the universe. Following Einstein, we shall call these statements Mach's principle.

Criticisms of Mach's Principle

Mach's ideas were received with considerable skepticism, even by philosophers who later favored the theory of relativity (which is partly based on Mach's ideas). Two main objections to Mach's principle can be distinguished in their writings, which we shall now examine.

Firstly, it was maintained that the laws of motion should be the same for all conceivable distributions of matter, so that the existence of centrifugal force cannot depend on whether or not there are any stars in the universe. Mach, on the other hand, had insisted that these laws have only been established in the existing universe which does contain stars, and that there is no need to suppose that they would be the same in a completely different one. On this point Bertrand Russell wrote as follows (*The Principles of Mathematics*, 1903):

> Mach has a very curious argument by which he attempts to refute the grounds in favour of absolute rotation. He remarks that, in the actual world, the earth rotates relative to the fixed stars, and that the universe is not given twice over in different shapes, but only once, and as we find it. Hence any argument that the rotation of the earth could be inferred if there were no heavenly bodies is futile. This argument contains the very essence of empiricism, in a sense in which empiricism is radically opposed to the philosophy advocated in the present work. The logical basis of the argument is that all propositions are essentially concerned with actual existents, not with entities which may or may not exist. For if, as has been held throughout our previous discussion, the whole dynamical world with its laws can be considered without regard to existence, then it can be no part of the meaning of these laws to assert that the matter to which they apply exists, and therefore they can be applied to universes which do not exist. Apart from general arguments, it is evident that the laws are so applied throughout rational dynamics, and that, in all exact calculations, the distribution of matter which is assumed is not that of the actual world. It seems impossible to deny significance to such calculations; and yet, if they have significance, if they contain propositions at all, whether true or false, then it can be no necessary part of their meaning to assert the existence of the matter to which they are applied. This being so, the universe is given, as an entity, not only twice, but as many times as there are possible distributions of matter, and Mach's argument falls to the ground.

A similar view was expressed by Cassirer, who wrote in 1910:

PLATE I. The sun's corona, photographed by the Royal Greenwich Observatory during the total eclipse of June 30, 1954. At such an eclipse the moon just blots out the normally visible disc of the sun, so that its faint outer extensions can be seen. If the sun and moon had very different angular diameters, such a complete picture of the corona could not be obtained.

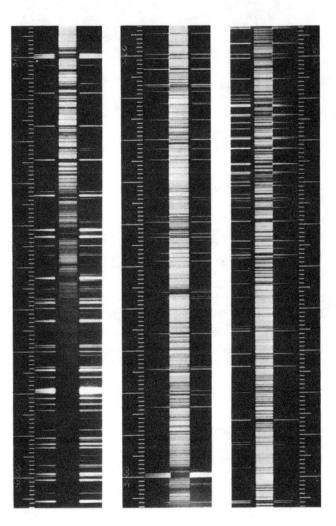

PLATE II. The spectrum of the sun between 3000 and 3300 angstroms (one angstrom is 10^{-8} centimeters), flanked by a laboratory spectrum of an iron source. The spectra of the stars have the same general appearance as the sun's spectrum and provide important clues to their physical properties.

PLATE III. The Small Magellanic Cloud. This photograph was taken on the 10-inch Metcalf triplet at the Boyden Station, Bloemfontein, South Africa.

PLATE IV. The Large Magellanic Cloud. This photograph was taken on the 10-inch Metcalf triplet at the Boyden Station, Bloemfontein, South Africa.

PLATE V. A mosaic photograph of the summer Milky Way, northern portion.

PLATE VI. A globular cluster of stars in Canes Venatici. About 100 globular clusters have been discovered, each containing about 1,000,000 stars.

PLATE VII. Clouds of stars in the region of Sagittarius. The center of the Milky Way lies behind these clouds at a distance of about 24,000 light-years.

PLATE VIII. An open cluster of stars (the Pleiades).

PLATE IX. The Andromeda nebula. It is a neighbor of the Milky Way and is probably similar to it in structure.

PLATE X.
Four
spiral nebulae.

PLATE XI. The main types of galaxies, as classified by Hubble. There are spirals (Sa, b, c), barred spirals (SBa, b, c), ellipticals (E 0, ... 7), and a few of irregular shape. The letters NGC and M refer to catalogues of celestial objects.

E0 NGC 3379 E2 NGC 221 (M32)

E5 NGC 4621 (M59) E7 NGC 3115

NGC 3034 (M82) NGC 4449

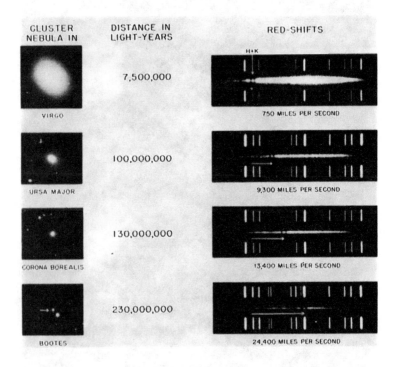

PLATE XII. The relation between red-shift and distance for galaxies. The horizontal arrows in the spectra indicate the shift of the H and K lines of calcium. The distances given in this plate should be multiplied by 10.

PLATE XIII. A pair of colliding galaxies. It is an extremely bright radio source.

PLATE XIV. The Foucault pendulum of the Franklin Institute, Philadelphia. Its supporting wire is 85 feet long, and the sphere weighs 1800 pounds. The pendulum is started in a north-south direction each day at 10 A.M. An hour later it is no longer swinging north and south, but about 10 degrees from that direction. On Newtonian ideas, this demonstrates the absolute rotation of the earth. According to Berkeley and Mach, it shows that the motion of the pendulum is governed by the stars.

The essential features [of a thought experiment] are separated from the accidental, which can vary arbitrarily without thereby affecting our real physical deduction. We need only apply these considerations to the discovery and expression of the principle of inertia in order to recognise that the real validity of this principle is not bound to any definite material system of reference. Even if we had found the law at first verified with respect to the fixed stars, there would be nothing to hinder us from freeing it from this condition by calling to mind that we can allow the original substratum to vary arbitrarily without the meaning and content of the law itself being thereby affected.

Secondly, Mach's principle was criticized because it gives no indication of the *nature* of the coupling between the stars and local matter which is supposed to give rise to inertial effects. Mach himself appears to have thought that the mere existence of the stars was enough, a suggestion which no doubt made the ghost of Euler repeat the words "very strange and contrary to the dogmas of metaphysics."

Presumably Mach came to this conclusion because the inertial properties of a body do not appear to depend at all on any intrinsic property of the stars; only their state of motion is needed to define inertial frames of reference. In particular, Newton's laws of motion and gravitation give an amazingly complete description of the behavior of the solar system, without any reference whatsoever to the physical properties of the stars.

This apparent irrelevance of the properties of the stars was perhaps the main argument used by those who attacked Mach. For instance, Bertrand Russell (unconsciously echoing Euler) wrote in 1927:

It is urged that for "absolute rotation," we may substitute "rotation relative to the fixed stars." This is formally correct, but the influence attributed to the fixed stars savours of astrology, and is scientifically incredible.

In the same vein Whitehead wrote in 1919:

> Surely this ascription of the centrifugal force on the earth's surface to the influence of Sirius is the last refuge of a theory in distress. The point is that the physical properties, size and distance of Sirius do not seem to matter.

And again in 1920:

> It is difficult to take seriously the suggestion that these domestic phenomena on the earth are due to the influence of the fixed stars. I cannot persuade myself to believe that a little star in its twinkling turned round Foucault's pendulum in the Paris Exhibition of 1851. Of course anything is believable when a definite physical connection has been demonstrated, for example the influence of sun-spots. Here all demonstration is lacking in the form of any coherent theory.

The Case for Mach's Principle

The objections described above are most easily disposed of in terms of a detailed theory of inertia which complies with Mach's principle. One such theory which answers these objections will be described in Chapter IX. Nevertheless, some sort of case should be made out for Mach's principle here, if only to supply a motive for constructing the theory in the first place.

I believe indeed that there is one overriding argument in favor of Mach's principle, an argument so compelling that it forces one to attempt to construct a theory in the hope that the objections we have mentioned can thereby be removed.

As an approach to this argument, let us call once again on Bertrand Russell—this time writing in defense of Mach (1924).

> The first to tackle the problem seriously was Mach. He considered that only observable phenomena should enter into the laws of an empirical science such as physics. Whether the earth rotates once a day from west to east,

as Copernicus taught, or the heavens revolve once a day from east to west, as his predecessors believed, the observable phenomena will be exactly the same. Nevertheless, Newtonian dynamics will explain Foucault's pendulum and the flattening of the earth at the poles if the earth rotates, but not if the heavens revolve. This shows a defect in Newtonian dynamics, since an empirical science ought not to contain a metaphysical assumption which can never be proved or disproved by observation—and no observations can distinguish the rotation of the earth from the revolution of the heavens. This philosophical principle, that distinctions which make no difference to observable phenomena must play no part in physics, has inspired a good deal of the work on relativity, and is advocated by many writers.

Russell is here objecting to the use of unobservables in physics. His point in essence is that unobservables are quite useless; they do not explain anything. This is illustrated in the following example. It is known that some metals lose all their electrical resistance when their temperature is lowered beyond a certain point. This phenomenon, known as superconductivity, has never been explained; indeed, it is one of the most famous unsolved problems of physics.[3] Now we could easily explain it in terms of unobservables. All we have to do is to assume that in these metals there sit invisible demons which push the current-carriers along in an appropriate manner. No one would take this theory seriously, of course. One reason for this, no doubt, would be the obviously *ad hoc* and, indeed, ludicrous appearance of the theory. But the fundamental reason for objecting to the theory is that the demons cannot be observed *except through the very phenomenon they were invented to explain.* The introduction of the demon thus adds nothing to what we know already. If, however, we could observe the demon *in some other way,* we would have discovered an important cor-

[3] Although a recent theory of Bardeen, Cooper, and Schrieffer looks very promising.

relation. In particular, we could verify the fact that just those metals which are observed to contain demons also become superconductors.

This points the contrast which exists between the hypothesis of absolute space and Mach's principle. According to Newton, the *only* way rotation relative to absolute space can be detected is from the existence of centrifugal and Coriolis forces. But absolute space was invented precisely in order to account for these forces. Like the superconducting demon, absolute space adds nothing to what we knew before. A dramatic example of this vacuousness occurs whenever a flywheel rotates too fast and centrifugal forces make it explode. To attribute this explosion to the action of an unobservable entity is like supposing with John Donne that bodiless angels can "molder a rock into . . . Atomes."

Contrast this with Mach's view that there is no such thing as *absolute* rotation, only rotation relative to other matter in the universe. In this view, centrifugal and Coriolis forces should act only on bodies which are rotating relative to the bulk of the matter in the universe—that is, relative to the fixed stars. These stars are then the physical agency responsible for the flywheel exploding. This is the type of correlation which the superconducting demon and absolute space both lack. We can now test whether a body which is not rotating as measured by a bucket of water or a Foucault pendulum is also not rotating relative to the fixed stars. This observational test of Mach's principle vindicates it to a high degree of precision. According to Newton, this agreement simply means that the system of stars is not, as it happens, rotating relative to absolute space. What for Newton is a remarkable fact is for Mach a causal relation.

The argument can be carried one stage further. For the agreement between the two methods of detecting non-rotation is not perfect. It was discovered in 1926 that the stars in the Milky Way are actually rotating around its center, the resulting centrifugal force making the Milky Way bulge out enormously. This means that a body not rotating relative to the stars in the Milky Way will also be acted on by a centrifugal force. In practice this force is negligible for a small body, since the rotation of the Milky Way is so slow (its period is about 100 million years). It is unfortunate that the Milky Way does not rotate much faster, for if it did this rotation would have been detected by Mach's time, and he would have been forced to postulate the existence of vast quantities of matter outside the Milky Way, relative to which it is rotating. In this way Mach could have used his principle to predict the existence of an extragalactic universe, a universe which was not discovered until fifty years later.

CHAPTER VIII

THE PRINCIPLE OF EQUIVALENCE

> Mr. Einstein's Theory of Relativity does not supersede the Newtonian law of Gravitation or of Inertia. It only says, Beware! The law of Inertia is not the simple ideal proposition you would like to make of it. It is a vast complexity. Gravitation is not one elemental uncouth force. It is a strange, infinitely complex, subtle aggregate of forces. And yet, however much it may waggle, a stone does fall to earth if you drop it.
>
> D. H. LAWRENCE

Introduction

At the end of the last chapter we committed ourselves to adopting Mach's principle, that the inertia of matter arises from the influence of distant stars. We are now faced with the task of devising a theory which incorporates this principle. Such a theory will be developed in this chapter and the next. The essential difficulty lies in discovering the *nature* of the interaction between nearby matter and distant stars which gives rise to the inertia of nearby matter. Only when this problem is solved will the main objection to Mach's principle be disposed of.

Inertia and Gravitation

This challenge was accepted by Einstein, who had been much impressed with Mach's arguments. By a stroke of genius he realized that the key to the solution lay in Galileo's experiment with falling bodies. This, of course, is one of the most famous experiments ever performed. Some historians deny that Galileo actually dropped anything from the Leaning Tower of Pisa, but we know that he conducted a thought-experiment which convinced him that a massive body and a light body would fall with the same acceleration, and so would land together (in the absence of air resistance). This experiment involves a heavy falling body. Split this body in imagination into two halves which fall side by side. Each of these portions will have the same acceleration as a body of just half the original mass. If the halves are combined again, they will presumably continue to have this acceleration. This would mean that a body has the same acceleration as another body half as massive. Galileo generalized this result, and concluded that bodies of different mass are equally accelerated by gravity. Very accurate measurements over a wide range of masses and materials have since confirmed his reasoning.

Galileo's result has an important consequence. According to Newton's second law of motion, the acceleration of a body is equal to the ratio of the force acting on it to its mass. Since falling bodies of *different* mass have the *same* acceleration, *the gravitational force acting on them must be proportional to their mass*. This property of gravitation is in striking contrast with the action of electric and magnetic forces, which do not induce the same accelerations in all bodies. An extreme example of this is provided by neutral bodies, which they do not accelerate at all. This difference between gravitation and electromagnetism was, of course, well known to all physicists since the time of Galileo, but no one saw its significance until 1907.

In that year Einstein recalled that there exists another type of force which, like gravitation, is proportional to the mass of the body it acts on. Einstein had in mind the inertial forces which we described in the last chapter. As we pointed out there, the strength of the inertial force acting on a body is proportional to the mass of that body (p. 93). In this respect, then, inertial forces are like gravitational forces, and unlike electric or magnetic ones.

Einstein realized that this similarity between gravitational and inertial forces makes it *impossible to distinguish between them.* This is best understood in terms of the way gravitational forces *can* be distinguished from electric or magnetic ones. Suppose we are told that there is a field of force present, and are asked to find out how much of it is gravitational and how much electrical. All we have to do is to measure the acceleration of one neutral body and one charged body. The acceleration of the neutral body tells us the strength of the *gravitational* part of the force. Now this part induces the same acceleration in the charged body and the neutral one. Thus the *difference* between their accelerations is a measure of the strength of the electric force.

Now suppose we tried to use the same technique to distinguish between a gravitational and an inertial force. Since both these forces are proportional to the mass of the body on which they act, we would be unable to make the required distinction. Whatever body we use to measure the force, the resulting acceleration will always be the same. We can determine the total strength of the force, but we cannot tell how much of it is gravitational and how much is inertial.

Einstein was fond of illustrating this state of affairs in the following way. Consider a man enclosed in a box somewhere in space far removed from gravitational forces. Suppose that the box is suddenly pulled by a rope, so that it accelerates relative to an inertial frame (Fig. 41a). The man inside the box may choose to consider himself as at rest throughout this experiment, but then the box represents a non-inertial frame of reference. Consequently an inertial force will be acting on the box. The existence of this inertial force will be obvious to the man: if he releases an object it will accelerate away from him (Fig. 41b). The crucial point is that *this acceleration will be the same for all the objects he releases,* since it is just equal and opposite to his own acceleration relative to an inertial frame. But this is exactly what would happen if, instead of

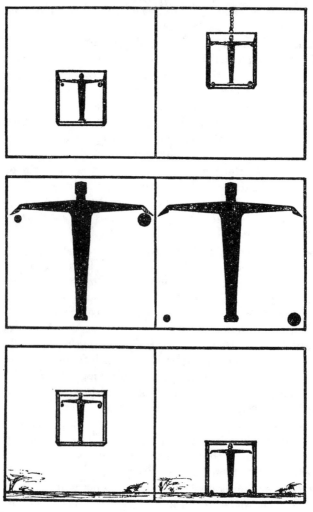

Fig. 41. (a) A man in a box out in space. The box is suddenly pulled upward by a rope. The man shoots upward with the box, but the balls near his hands remain where

being pulled by a rope, the box were acted on by a gravitational force (Fig. 41c). This means that the man will not be able to tell which of these two possibilities is the correct one!

So far this conclusion is based entirely on the similarity in the observed response of *massive* bodies to a gravitational force. The possibility clearly arises that there is some other criterion by which an inertial force can be distinguished from a gravitational one—for instance, from the behavior of light, or some subtle atomic phenomenon on a microscopic level. What Einstein did was to elevate Galileo's experiment into a principle—the famous principle of equivalence. According to this principle, *there is no criterion whatsoever by means of which an inertial force can be distinguished from a gravitational one.*

Einstein's Explanation of the Principle of Equivalence

This principle of equivalence is a plausible generalization of Galileo's discovery. At the same time it is very puzzling. Why should inertial forces mimic gravitational ones so closely that they can never be distinguished as definitely inertial? Einstein answered this question in a beautifully simple way by saying that *inertial forces are themselves gravitational in origin.* Now gravitational forces must originate from something —they must have sources in the form of lumps of matter. Which lumps of matter, then, are the sources of inertial forces?

---

they were. (b) The same situation from the man's point of view. He experiences an inertial force acting downward. This force also acts on the balls and induces the same acceleration in each of them although they have different masses. This acceleration is equal and opposite to the man's acceleration in a. (c) A different situation which seems the same to the man. He is first falling toward a gravitating body and is then brought to rest at its surface, while the balls continue to fall with the same acceleration.

With Mach's principle in mind, Einstein had no difficulty in answering this question. Indeed, the difficulty was really the other way around—Mach had provided the sources of inertial forces but had left the nature of these forces quite obscure. All Einstein had to do was to put the two problems together. He therefore concluded that the inertial forces which arise in a non-inertial frame of reference are gravitational forces exerted by the stars.

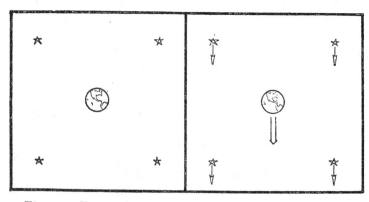

Fig. 42. *Einstein's suggested relation between inertial forces and the gravitational forces of the stars. (a) If the stars are unaccelerated, their gravitational forces on the earth cancel out by symmetry. According to Einstein, there are then no inertial forces acting on the earth. (b) When the stars accelerate, they exert a net gravitational force on the earth which is regarded as an inertial force.*

These gravitational forces presumably cancel one another out in frames of reference at rest relative to the stars, if the stars (or rather galaxies) are distributed more or less symmetrically (Fig. 42a). There will then be no inertial forces in such frames—that is, they must be inertial frames. On the other hand, we know that there are inertial forces acting in frames which accelerate relative to the stars. If Einstein is right, this means that the gravitational force of the stars must arise from their *acceleration* relative to the non-inertial frames (Fig. 42b). In other words, *an accelerating star must exert a different gravitational force from a non-accelerating one.* This fundamental requirement is incorporated in the detailed theory described in the next chapter.

Einstein's explanation of the principle of equivalence gives rise to many questions, such as: Are there enough stars in the universe? In other words, when they accelerate, is their combined gravitational effect large enough to give rise to inertial forces of the observed magnitude? Some people, indeed, maintained that there were not enough stars. Eddington, for instance, wrote in 1920:

> If the earth is non-rotating, the stars must be going round it with terrific speed. May they not in virtue of their high velocities produce gravitationally a sensible field of force on the earth, which we recognize as the centrifugal force? This would be a genuine elimination of absolute rotation, attributing all effects indifferently to the rotation of the earth, the stars being at rest, or to the revolution of the stars, the earth being at rest; nothing matters except the relative rotation. I doubt whether anyone will persuade himself that the stars have anything to do with the phenomenon. We do not believe that if the heavenly bodies were all annihilated it would upset the gyrocompass. In any case, precise calculation shows that the centrifugal forces could not be produced by the motions of the stars, so far as they are known.

Similarly, the English mathematician Sir Edmund Whittaker wrote as recently as 1953:

> When we confront this hypothesis with the facts of observation, however, it seems that the masses of whose existence we know—the solar system, stars, and nebulae—are insufficient to confer on terrestrial bodies the inertial mass that they actually possess: and therefore if Mach's principle were adopted, it would be necessary to postu-

late the existence of enormous quantities of matter in the universe which have not been detected by astronomical observation, and which are called into being simply in order to account for inertia in other bodies. This is, after all, no better than regarding some parts of inertia as intrinsic.

In the next chapter we shall come to the conclusion that Eddington and Whittaker were mistaken, and that there is enough matter in the universe to account for the whole of local inertia. Indeed, the fact that there is exactly the right amount of matter, neither too little nor too much, will turn out to be no accident but of great significance.

CHAPTER IX

THE ORIGIN OF INERTIA

> And fliers and pursuers
> Were mingled in a mass.
> MACAULAY

Introduction

In order to test Einstein's idea that inertial forces are actually gravitational forces exerted by *accelerating* stars, we need a theory which tells us how much gravitation is produced by a moving star. It was this need that led Einstein to devise his general theory of relativity, which was published in 1915. (It should not be confused with his special theory of 1905, which has nothing to do with gravitation.) Unfortunately, although general relativity is based on simple physical principles, it is very involved mathematically—so involved, indeed, that the extent to which it incorporates Mach's principle is still a matter of controversy. Because of this complication a description of Einstein's theory is deferred to a later chapter.

Nevertheless, this need not prevent us from pursuing the problem of Mach's principle. It is possible to construct a simplified version of Einstein's theory which will still tell us how much gravitation is produced by an accelerating star, but which is sufficiently simple for its implications about inertia to be completely worked out. The full complexity of Einstein's theory is needed for some problems but fortunately not for this one.

Simple Theory of Gravitation

Our task is to construct a theory from which we can calculate the gravitational force produced by any body. The first possibility that comes to mind is simply to use Newton's law of gravitation, which asserts that the gravitational force exerted by a body decreases inversely as the square of its distance—the famous inverse square law. However, we also need to know the force exerted by a *moving* body, and this complication was not considered by Newton. He seems to have taken it for granted that the gravitational force exerted by a body does not depend on its state of motion. But we require a stationary and an accelerated body to exert different forces, if the accelerated motion of the stars is to be held responsible for inertial forces.

This is not the first occasion on which the forces exerted by stationary and moving bodies have been distinguished. In 1835, Karl Friedrich Gauss, (1777–1855), a German mathematician and astronomer, discovered that a moving charge exerts a different electric force from a charge at rest. In accordance with his usual custom, he did not publish this result, and it was rediscovered in 1846 by his colleague, the German physicist Wilhelm Weber (1804–91). It has since been enshrined in Maxwell's theory of electromagnetism (1865). Of particular interest to us is the force exerted by an *accelerating* body. In the electrical case this force differs from the inverse square (Coulomb) force of a stationary charge in a characteristic way, which has the important consequence that when a charge accelerates it emits electromagnetic waves. Maxwell calculated the velocity of these waves and found it to be the same as the observed velocity of light. This result led him to his greatest discovery, that light itself consists of electromagnetic waves.

This triumph encouraged the French astronomer F. Tisserand to apply these ideas to gravitation. In 1872 he suggested that the gravitational force exerted by a moving body might obey the same laws as the electric and magnetic forces exerted by a moving charge. Tisserand did not have in mind the problem of inertia, and the only interesting result he obtained was that the planets would deviate slightly from their Newtonian orbits around the sun. Such a deviation had in fact been detected in the motion of Mercury by the French astronomer Leverrier in 1845, but Tisserand was able to reproduce only a fraction of this deviation theoretically. The motion of Mercury remained a mystery until 1915, when Einstein showed that it obeyed his new laws of gravitation.

The motion of Mercury is one of the problems which needs for its solution the full complexity of Einstein's theory. Fortunately this is not true of the problem of inertia. We shall, therefore, revive Tisserand's idea and suppose that the gravitational force exerted by a moving body obeys the same laws as the electric and magnetic forces exerted by a moving charge.

We are now in a position to calculate the gravitational force exerted by the stars when they accelerate. If our theory is sound, this force will be just equal to the inertial force postulated by Newton. Now, as we saw on page 93, when a body is regarded as being at rest the inertial force acting on it is equal to its mass times its absolute acceleration, or rather from our present Machian point of view, its mass times the acceleration of the stars. Does Tisserand's theory of gravitation lead to this result?

To answer this question we must first determine the force exerted by *one* star; then we can add together the contributions of all the stars. Since the force exerted by a star is assumed to obey the same laws as the force exerted by a charge, we shall begin with a detailed description of electrical forces.

If a charge is stationary the force it exerts on another charge is proportional to each of the charges and inversely proportional to the square of the distance between them (Fig. 43a). This is the famous law named after Charles Coulomb, a French physicist (1736–1806), and is the analogue of Newton's inverse square law for gravitation.[1] At this stage, then, we have not gone beyond Newton's theory. It is only when we go on to consider the additional force exerted when the source charge *accelerates,* and its analogue for gravitation, that we strike new ground.

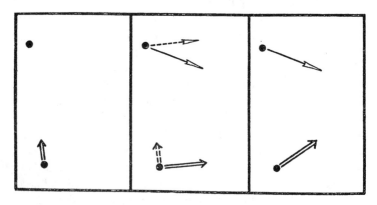

Fig. 43. *The forces exerted by an electric charge. (a) A stationary charge exerts a force on another charge that is (i) radial and (ii) proportional to the inverse square of their distance apart. (b) An accelerating charge exerts an additional force that is (i) transverse, (ii) proportional to the inverse first power of the distance, (iii) proportional to the transverse component of the acceleration. (c) The total force exerted by an accelerating charge.*

This additional force is like the Coulomb force in that it is proportional to each of the charges, but there the resemblance ends (Fig. 43b). There are two major differences and two minor ones. The first major difference is that the force decreases inversely with the distance instead of with the square of the distance—doubling the distance only halves the force instead of quartering it. The second major difference is that the force is proportional to the acceleration of the source

[1] Except that like charges *repel* one another.

THE ORIGIN OF INERTIA

charge (so that this force is zero when the source has no acceleration). These two new features are of fundamental importance for what follows. The two minor features are shown in Fig. 43b; we need not describe them here.

These two forces—the Coulomb force and the acceleration force—can be taken over directly into gravitation. The only change needed is to replace electric charge by "gravitational charge." We have learned from Newton to call it gravitational *mass*, but it can hardly be overemphasized that this expression must not be confused with inertial mass, particularly as we intend to *derive* this latter concept from our theory of gravitation. With this warning, then, we shall use the phrase "gravitational mass" to refer to the gravitational analogue of charge.

The Gravitational Force Exerted by Accelerating Stars

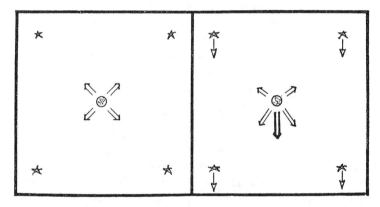

Fig. 44. The relation between inertial forces and the (charge-like) gravitational forces of the stars. (a) If the stars are unaccelerated, their gravitational forces on the earth cancel out by symmetry. No inertial forces are then acting (cf. Fig. 42a). (b) When the stars accelerate, they exert a net gravitational force on the earth, which is regarded as an inertial force (cf. Fig. 42b).

We come at last to our main task, that of calculating the total gravitational force exerted on a body by the stars when they accelerate. Now the inverse square parts of their forces cancel out by symmetry, since there is presumably the same number of stars in every direction (Fig. 44a). In contrast, the acceleration forces do not cancel out; there is a net force on the body which acts in the same direction as that of the stars' acceleration (Fig. 44b).

This net force will be proportional to the acceleration of the stars, since this is so for each individual star. This means that we have already achieved half of our task, which was to make the gravitational force of the stars equal to their acceleration times the inertial mass of the body. Our remaining task is to see whether the inertial mass of the body is equal to all the other factors besides acceleration which determine the total force of the stars.

These factors depend on the gravitational masses of the stars and on their distances. We can assume for simplicity that all the stars have the same gravitational mass, but we cannot ignore the differences in their distances. This raises a serious problem: how can we combine the forces exerted by the stars unless we know all their individual distances?

The way to approach this problem is to discover which stars make the main contribution—as a first approximation we can concentrate on them. Now we faced a similar problem earlier in this book. In order to resolve Olbers' paradox, we began by locating the stars which in his calculation made the main contribution to the light of the night sky. We found that these stars are at very great distances. The reason for this is that, although the contribution of a star decreases inversely as the square of its distance, the number of stars in a spherical layer *increases directly* as the square of the distance. Each spherical layer of stars will thus make the same contribution, except for layers at very great distances, some of whose stars lie partly behind nearer ones. This means that the main contribution comes from distant stars, since there are many more layers at large distances than at small ones. Indeed, Olbers would have obtained an infinite answer but for the shielding of very distant stars by nearer ones.

The dependence on distant sources is even more striking in our gravitational problem. In the first place, the gravitational force that arises from a star's acceleration decreases inversely as the distance rather than as the square of its distance. Since the number of stars in a spherical layer increases as the square of the distance, the contribution from a spherical layer is now *greater* for more distant layers; indeed, it is proportional to the layer's distance. Furthermore, there are now no shielding effects to keep the total from being infinite. This "gravitational Olbers' paradox" is solved in the same way as the original paradox, namely by taking into account the recession of distant sources. One can show that there is a gravitational Doppler effect which reduces the contribution from very distant regions and keeps the total finite.

This gravitational Doppler effect clearly reduces to some extent the influence of the most distant sources, but although it succeeds in making the total gravitational force of the stars finite, the main contribution still comes from very distant stars. For instance, 80 per cent of the total force is exerted by sources too distant to be detected by the 200-inch telescope. It is this result which enables us to add the contributions of all the stars, without having to know their individual distances. For we can approximate to the distribution of the stars by imagining that their material is spread uniformly throughout space. All we really need to know is the gravitational mass per unit volume, or gravitational density of this uniform distribution. On the other hand, had the most important stars turned out to be a few nearby ones, we should have had to take into account any irregularities in their distribution—for instance, an exceptionally near star would have had an important influence on the result. But because by far the major contribution comes from extremely distant stars, their average "smeared out" contribution is an excellent approximation to the total force. A few irregularly placed stars will have only a small influence on the result.

The total gravitational force exerted by the stars depends, then, on the average gravitational density of matter at great distances, which we shall call ρ_G. It depends, in addition, on another quantity (besides their acceleration), namely the rate of expansion of the universe. This rate is relevant since it determines the size of the gravitational Doppler effect. It is measured by the value of Hubble's constant, which we shall call τ.

Finally, the force depends on the gravitational mass, m_G of the body itself. Calculation shows that the value of the force is actually $\rho_G \tau^2 m_G$ times the acceleration of the stars.[2] Now our aim was to make this force equal to the *inertial* mass (m_i) of the body times the acceleration of the stars. Thus we have derived inertial mass from our theory *if* it can be identified with the factors in our formula which multiply the acceleration. In other words, we are going to try to define inertial mass by the formula:

$$m_i = \rho_G \tau^2 m_G$$

We must now unpack the meaning of this strange-looking equation. Does the quantity m_i which is defined by it have all the properties of an inertial mass? It obviously enters the law of motion in the right way—that is, its product with acceleration gives the inertial force—indeed, this was precisely what led to our definition. In addition this definition shows that the inertial and the gravitational mass of a body (m_i and m_G) are proportional to one another. We have already seen (p. 108) that this is one of the basic properties of gravitation, but whereas Newton had to *assume* it, we have been able to *derive* it.

The Gravitational Constant

We can go even further than this, for our theory also tells us the value of the constant of proportionality between m_i and

---

[2] For simplicity, I am leaving out a numerical factor which does not differ much from unity.

m_G. To see what this implies, we shall first recall the Newtonian attitude to this quantity. According to Newton, the gravitational force between two bodies is proportional to the product of their inertial masses,[3] the constant of proportionality being known as the gravitational constant G. This constant is a measure of the *strength* of gravitation—its value determines how much gravitation is produced by a body of given inertial mass. It is generally considered to be one of the fundamental constants of nature, and it plays a vital part in our theory of Mach's principle.

The first accurate measurement of G was made by the English physicist Henry Cavendish in 1798. He succeeded in determining the force of attraction between two spheres of known inertial mass and separation. This measurement is a difficult one to make, as the force is very small. Of course gravitation seems large to us, but compare the mass of the earth with the mass of a small sphere! Some physicists have, in fact, used the attraction of the earth to determine G, but unfortunately its mass is not known very accurately. The smallness of gravitation is reflected in the smallness of the value of G: in units in which lengths are measured in centimeters and masses in grams, it has the value 6.6×10^{-8}. This means that the gravitational force between two bodies of mass one gram which are one centimeter apart accelerates them by less than one ten-millionth of a centimeter per second per second. It also means that the electrical force between two electrons is 10^{42} times greater than the gravitational force.

We are emphasizing the significance of the gravitational constant because it plays a key role in our theory of Mach's principle. Since this theory prescribes a definite value for the ratio of m_G to m_i in terms of the properties of the stars, it also determines the value of G in terms of these properties. To obtain this relation it is convenient to replace the gravitational density ρ_G of the stars by their inertial density ρ_i using, of

---

[3] And, of course, to the inverse square of their distance apart.

course, the same constant of proportionality as for m_G and m_i. The final result is:
$$G\rho_i \tau^2 = 1$$
This relation is the culmination of our whole long discussion. It brings together three quantities which at first sight are completely independent of one another. Like most relations of this sort, it can be exploited in many different ways. Let us examine a few of these.

First of all, we can use the relation to calculate ρ_i, the density of matter at great distances. To do this we do not require any information obtained by telescopes, for an approximate value of Hubble's constant τ can be deduced from the amount of light in the night sky (p. 80). When this is combined with the Cavendish value of G we obtain an average density corresponding to about 1 hydrogen atom in every 10 liters of space. More striking than the actual value is the obtaining of a definite result at all. It means that from observations restricted to our own neighborhood we can deduce an approximate value for both Hubble's constant and the average density of matter at great distances.

Our ability to calculate this average density is not surprising, once we understand the significance of the relation between the gravitational constant G and the amount of matter in the stars. This significance can be expressed as follows. We saw when we defined the gravitational constant G (p. 123) that it is a measure of the gravitational force produced by a body of given inertial mass. We can reverse this, and say that G, or rather its reciprocal, is a measure of the inertial mass of a body which produces a given gravitational force. Now in the present theory a body produces the same gravitational force whatever other bodies there are in the universe. On the other hand, its inertial mass is induced into it by all these other bodies. Hence G, which measures the ratio of inertial mass to gravitational mass, is determined by these bodies. Our formula for G, the density of matter, and Hubble's constant is just the mathematical expression of this physical relationship.

This result suggests that Eddington and Whittaker were

probably mistaken in thinking that there is not enough matter in the universe to account for the whole of inertia. Indeed, there is a good reason why the universe should contain precisely the right amount of matter, for this amount is simply reflected in the value of G—had it contained more matter, for instance, G would have been still smaller than it actually is.

This determination of the value of G adds considerably to the fundamental significance of Mach's principle. For this principle has enabled us to obtain a *quantitative* result, namely to account for the observed value of the gravitational constant. In contrast, the Newtonian theory of absolute space does not determine this constant, since it does not specify how much inertia is conferred on matter by absolute space.

Our quantitative result is also important in that it explains the *apparent* irrelevance of the properties of the stars to the inertial behavior of matter—the point which, it may be remembered, was mainly responsible for the outspoken criticisms that Mach's principle received. For we see now that the universe makes itself felt in local phenomena at just the two points where the Newtonian scheme contains arbitrary elements, namely in the choice of inertial frames and in the value of the gravitational constant.

Centrifugal and Coriolis Forces

Our theory must also be able to account for the centrifugal and Coriolis forces that act on a rotating body. If we regard such a body as at rest, then it is the stars that are rotating around it. The gravitational forces exerted by these rotating stars must be just the centrifugal and Coriolis forces. Is this requirement satisfied?

Now, according to our theory, these gravitational forces will be similar to the electric and magnetic forces exerted by a rotating system of charges. The electric-type force of such a system is zero at its center but not at other points: indeed, it behaves in the required way to be a centrifugal force—that is, it acts radially away from the axis of rotation.

Furthermore, since a magnetic force acts only on a *moving* charge, its gravitational analogue acts only on a moving body

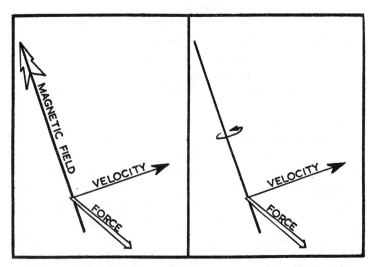

Fig. 45. The similarity between magnetic and Coriolis forces. (a) The magnetic force on a moving charge. This force acts at right angles to the magnetic field and to the direction of the charge's velocity. (b) The Coriolis force acts in the same way as a magnetic force, the axis and angular velocity of the rotating frame of reference replacing the magnetic field.

(Fig. 45). This is just the characteristic of the Coriolis force that we discussed on page 91. Moreover, both forces act transversely to the velocity, and in opposite directions for oppositely directed velocities.[4]

We can thus identify centrifugal and Coriolis forces with the

[4] The forces involve the vector product of the velocity and the magnetic field in one case, and the velocity and angular velocity of the rotating frame in the other.

"grav-electric" and "gravo-magnetic" forces exerted by the stars when they rotate. I should emphasize that this use of the words "grav-electric" and "gravo-magnetic" is only by way of analogy, in the sense that these forces obey the same laws as electric and magnetic forces. They are in our theory completely gravitational in origin and significance.

Needless to say, we require these forces to have the right strength before they can be identified with centrifugal and Coriolis forces. The condition that there are precisely enough stars for this to be so is just the relation that we had before, namely $G\rho\tau^2 = 1$. This helps to confirm the idea that centrifugal and Coriolis forces arise from the gravitational effects of a rotating universe.

General Conclusions

It is instructive, now, to stand a little away from the theory and to see what general conclusions can be drawn from it. Its essential feature is that it is the very distant stars which make the main contribution to inertia. In fact 80 per cent of the inertia of local matter arises from the influence of galaxies too distant to be detected by the 200-inch telescope. This means that our theoretical value for the average density of matter must relate to very distant regions of the universe. Since our value for this density is much the same as that derived from direct observation, it appears that the regions of the universe lying beyond the reach of the 200-inch telescope may not be very different from those already observed.

We have here theoretical support for Hubble's belief that in his receding galaxies he had observed a typical sample of the universe as a whole. Further support for this belief comes from the resolution of Olbers' paradox, which implies that the unsurveyed part of the universe must also be expanding, and at a rate similar to that derived by Hubble.

This shows how useful are those local phenomena which are strongly influenced by distant regions of the universe. For if we possess a theory of this influence we can use our local knowledge to discover something of the behavior of matter at great distances. Such theories, like telescopes, are tools for exploring remote regions of the universe.

The idea that distant matter can sometimes have far more influence than nearby matter may be an unfamiliar one. To make it more concrete, we give a numerical estimate of the influence of nearby objects in determining the inertia of bodies on the earth: of this inertia, the whole of the Milky Way contributes only one ten-millionth, the sun one hundred-millionth, and the earth itself one thousand-millionth!

This shows very clearly how unimportant for inertia are those parts of the universe with which we are most familiar. If this had been realized when the location of the spiral nebulae was in dispute, it could have been used to settle the dispute by showing that the Milky Way is an insignificant inhabitant of the universe.

An unfortunate consequence of this unimportance of nearby matter is that we cannot appreciably alter the inertia of a body by changing its environment. We are thus unable to use the physicist's normal experimental method of discovering influences. As a result we can easily be misled, as Newton was, into thinking that inertia is an intrinsic property of matter—one that belongs to itself alone, since it does not appear to depend in any way on its environment. This shows how difficult it will be to decide whether any apparently constant property of matter is indeed intrinsic, or whether it arises as a result of the influence of distant matter.

To see what types of influence may be at work, it is useful to have a table showing, for various possible laws of force, what is the relative importance of near and distant matter. In the following table we list in the first column the way in which each force decreases with distance. In the second column we give the fraction of the total force exerted by the stars at unit distance, on the assumption that the stars are equally spaced at this unit distance throughout the universe. Finally, in the third column, we give the extra force which would be exerted

THE ORIGIN OF INERTIA

by an exceptional star, nearer than the unit distance by a factor of a million.

If we take a point at random in space, not especially near any one star, the second column estimates what fraction of the total force acting there is exerted by the nearest stars. For those special points whose distance from the nearest star is only one millionth of the unit distance, the third column gives the nearest star's additional contribution to the total force.

Force decreases with distance as $\dfrac{1}{r^u}$	Fraction of total force which is exerted by nearest neighbors at unit distance	Extra force which is exerted by a very close neighbor at a distance of 10^{-6} of unit distance
4	½	10^{23}
3	1/10	10^{16}
2	10^{-4}	10^{7}
1	10^{-12}	10^{-8}

Total force at a typical point $= 1$

The second column of the table shows that the switch in importance between near and distant matter occurs somewhere between an inverse cube and an inverse square law of force. If there is an unusually near star, the third column shows that it still dominates for an inverse square law. Since this is the law that applies in Olbers' paradox, we can understand why it is so much lighter during the day than it is at night. For the sun is a specially near star, and its light is far greater than the total amount of light from all other stars combined.

By the time we reach the inverse first power law, which characterizes our theory of inertia, even a specially near star makes a negligible contribution—only one part in a hundred million. If such a law governs other processes besides gravitation, many of the apparently intrinsic properties of matter must really arise from its interaction with the rest of the universe. Indeed, we cannot at the present time set any limits to the extent of the universe's influence on local phenomena. It is a problem for the future to discover where these limits lie.

CHAPTER X

THE CLOCK PARADOX

For I dipt into the future, far as human eye could see,
Saw the Vision of the world, and all the wonder that would be.
                                        TENNYSON

Introduction

In the last two chapters we have explored the gravitational link which connects us to the universe as a whole, and revealed its fundamental, but well-disguised, role in Newton's second law of motion. It might be thought that this role would exhaust the gravitational influence of distant matter. This is by no means so, for its influence can be felt in many ways. We shall see, for instance, that it enables us to resolve a paradox which has been the source of much confusion—I mean the so-called "clock paradox." This paradox was devised by Einstein back in 1905 but it still arouses controversy.

The Paradox

A man leaves the earth in a rocket ship, flies to a distant point, and then returns to the earth. According to Einstein, the man ages less during this journey than the people who remain on earth. Stated more abstractly, this means that a clock in the rocket ship registers a shorter time for the duration of the journey than does a clock on the earth. By way of illustration, let us suppose that the rocket's speed is one seventh of

1 per cent less than the speed of light, and that when it returns people on the earth are twenty years older than when it left. Then Einstein showed that the man himself will have aged only one year!

Now that space travel seems imminent, such a journey sounds more practicable than it did in 1905. It is no wonder, then, that there has been a revival of interest in Einstein's result. It can, in fact, be made even more startling: if the rocket were to fly yet closer to the speed of light, the man could return still a youth, to find his remote descendants peopling the earth. Here is time travel indeed! Who would not rejoice at the possibility of sharing man's understanding of the universe five hundred or a thousand years hence—or even of knowing whether he has blown himself up by then? *But this is not yet the paradox.*

To reach it we must go one step further. As Einstein pointed out, and as has been emphasized repeatedly in this book, all motion is relative. This means that we can regard the man in the rocket as at rest throughout the journey, while the earth and its inhabitants shoot out and then return. But in that case we should expect the earthbound people to age less than the rocket man! Here, then, is the paradox: according to relativity, a "moving" man ages less than a "stationary" one, but also according to relativity, either man may be regarded as the "moving" one.

As a first step toward the resolution of this paradox, let us see with Einstein why the rocket man ages less when he is regarded as the one who moves out and returns. After this we shall examine what happens when the rocket man is considered to be stationary. To simplify the problem consider just two men, A and B. A is the stay-at-home, while B moves out in the rocket ship and then returns. Let us suppose that each of them carries a source of light which emits 50 waves per second in the other's direction.[1] By comparing what each of

[1] This frequency does not actually correspond to visible light, but I want to keep the numbers simple.

THE CLOCK PARADOX

them sees of the other's light, we can, in fact, compare the amounts by which they age during the journey.

As the rocket man B moves out he will see A's light Doppler-shifted toward the *red*—that is, he will receive *fewer* than 50 waves per second. Let us suppose that he receives 49 waves per second (the actual number depends on his speed). When B turns around to begin the return journey he will, *at that instant*, see A's light change from being red-shifted to being blue-shifted—that is as soon as he starts to move toward A he will receive more than 50 waves per second. If he returns at the same speed as he went out, he will in fact receive 51 waves per second. Thus B sees red for half of his journey and blue for the other half. If each half of the journey takes, according to B, 100 seconds, then he receives altogether 10,000 waves ($51 \times 100 + 49 \times 100$).

Compare this with what the stay-at-home A sees of B's light. As B recedes A will see the same red-shift as B did on his outward journey—that is, A will receive 49 waves per second. But—and here is the crucial point—when B turns around and starts for home A *will still see red for a little while*. The reason for this is that at the moment of turn-around A is receiving light emitted by B *before* the turn-around (as the light takes time to reach A) (Fig. 46). Indeed, when A first sees a blue-shift B will have performed more than half of his journey and will be on his way home. This means that altogether A sees blue for a shorter time than he sees red—that is, waves arrive at the rate of 51 per second for a shorter time than they arrive at 49 per second.

How many waves does A receive altogether? If the whole journey lasts 200 seconds for A as well as for B, A will receive *fewer* than 10,000 waves, since the blue-shift does not last long enough to compensate for the red-shift. On the other hand, B emits 50 waves per second for 200 seconds—that is, 10,000 waves—*none of which can get lost*. So A *must* receive these 10,000 waves by the time B reaches him, since B cannot overtake a wave moving with the speed of light. This is only possible if A receives waves for *longer* than 200 seconds. In other words, the journey lasts longer for A than it does for B; A has aged more than B. A detailed calculation, based on the formula for the Doppler shift, leads to the figures quoted at the beginning, namely twenty years as against one year for a speed of one seventh of 1 per cent less than the speed of light.

This, in outline, is the argument used by Einstein to show

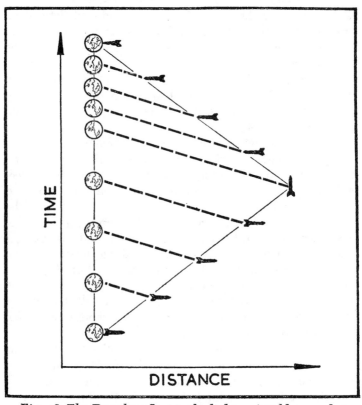

Fig. 46. The Doppler effect in the light emitted by a rocket ship which recedes and then returns. Light of lowered frequency (that is, reddened light) is received for more than half the duration of the journey (as measured by clocks on the earth).

that the moving man will age less than the stay-at-home. It is not surprising that it results in the men aging differently, since they are treated differently in the calculation. Whereas B is assumed to see the light change color as soon as he changes his velocity—that is, at half time—A sees a color change *after* half time. This seems reasonable in terms of the diagram of Fig. 46. Nevertheless, the paradox is still with us, since we can always describe the journey by saying that B remains at rest while A moves out and returns. In that case we might expect Fig. 46 to apply with A and B interchanged. Why does it not? What is the difference between A and B?

Resolution of the Paradox

The difference between A and B is that B is a man in a rocket—that is to say, he needs a powerful fuel to drive him away from A; whereas, if we regard A as the man who moves, he still does not need the help of any such force. In other words, the difference between A and B is that B accelerates *relative to an inertial frame* when the rocket fuel is burning, while throughout the journey A remains unaccelerated relative to an inertial frame.

We are reminded by this distinction of Newton's experiment with the rotating bucket of water (p. 94). There, it will be remembered, Newton could detect rotation of the water just by looking at its surface. In other words, he could tell whether an object was accelerating relative to an inertial frame just by inspecting the object. Here, too, we can tell which man has accelerated relative to an inertial frame just by inspecting their ages.

But we can go further than this. Our discussion of Newton's experiment showed that "rotation relative to an inertial frame" actually means "rotation relative to distant matter." If we use this idea in the present problem we can say that the difference between the two men is that one of them has accelerated relative to distant matter while the other has not.

This is undoubtedly a genuine difference between the two men. But does it resolve the paradox? Can acceleration relative to distant matter change the rate at which a man ages? The best way to answer this question is to take the point of view that the rocket man B is at rest. Now a naïve application of our previous calculation would suggest that the one who ages less is A. Have we forgotten something? We have indeed: for when B is considered to be at rest, not only will A be accelerating, *but so also will distant matter*. This acceleration of distant matter produces gravitational effects. Presumably one of these effects is to influence the rate at which A and B age, and thereby to swing A around from aging less than B to aging more. To test this idea, we must consider how gravitation affects the rate at which a man ages or, more abstractly, how it affects the rate of a clock.

The Effect of Gravitation on a Clock

Consider an atom sitting on the surface of the sun and emitting light in the direction of the earth (whose gravitational influence we can here ignore). We can use this light as a clock, by counting its waves instead of swings of a pendulum or rotations of a clock hand. Time will then be measured in waves instead of seconds. The advantage of this is that we can use the known effect of the sun's gravitation on the light to tell us its effect on the light's performance as a clock. In this indirect way we can establish the effect of gravitation on the rate of *any* clock, since the principle of equivalence assures us that gravitation affects all clocks equally.

Now the light reaching us from an atom on the sun loses energy in moving against the sun's gravitational force, just as a ball thrown up from the earth loses energy. But whereas the ball slows down, the energy loss in the light is manifested by a decrease in its frequency—a red-shift.[2] An observer on the

[2] This is obvious in the photon picture of light, for the energy of a photon is proportional to the frequency of the light it represents.

earth, comparing this reduced frequency with the frequency of a similar atom in his neighborhood, will judge the solar clock to be going *slow*. A time interval of, say, 30 waves as measured by a terrestrial clock will last less than that according to the solar clock. In other words, because of the sun's gravitation[3] the rate of a clock on the sun is slower than the rate of a similar clock on the earth. *Gravitation makes a clock tick slower.*

Equipped with this result, we can now test whether the gravitational effect of distant matter will swing A around from aging less than B to aging more than B. Since we are here supposing that B remains at rest throughout the journey, distant matter will accelerate on three occasions: first, at the beginning of the journey; secondly, when A turns around; and thirdly, at the end when A comes to rest. On each of these occasions distant matter will produce gravitational effects. Now B manages to remain at rest despite these gravitational effects, because on each occasion they are exactly compensated by the thrust of the rocket. On the other hand, these gravitational forces affect A and account for his various accelerations and decelerations. Thus we have a consistent description of the journey from B's point of view—if we can swing A's aging around.

At the beginning and end of the journey A and B are more or less in the same place, so that their clocks will be equally affected by gravitation. But when A is forced to turn around in the middle of the journey A and B are far apart. Gravitation now affects their clocks unequally, just as it does with a solar and a terrestrial clock. During the time the gravitational force is acting, A's clock ticks *faster* than B's. As a result their clocks will register different times for the duration of the turn-around —that is, for the duration of the gravitational force. In fact this turn-around lasts *longer* for A than it does for B. This speeding up of A's clock acts in the opposite direction to the slowing down we naïvely expected, and so is in the right direc-

[3] The rate of a clock depends on the gravitational *potential* it is in, not on the force.

tion to produce the desired swing. A precise calculation shows that indeed it just suffices: A ages more than B by precisely the amount which we calculated on the assumption that the rocket man B is moving.[4]

This disposes of the paradox. Either man may be regarded as stationary; the difference between them springs from their different relations to distant matter. The clock paradox thus has exactly the same status as Newton's experiment with the rotating bucket of water. Some people have claimed that the two men are, in fact, symmetrical, and so should age equally during the journey. They have clearly overlooked the relevance of distant matter. If there were no such matter the men would indeed be symmetrical and would age equally, but in that case there would be no such thing as inertial frames or inertia, and no rocket would be needed to accelerate B. Life would be quite different in such a universe.

There is no doubt that in the actual universe the rocket man ages less than the stay-at-home. It is tempting to exploit this effect in order to "dip into the future," but, unfortunately, with existing rockets one would run into serious fuel problems. If these problems can be solved time travel will become a reality —but only travel into the future. The past will still be inaccessible.

[4] Mathematically minded readers may amuse themselves by proving this.

CHAPTER XI

THE GENERAL THEORY OF RELATIVITY

> A man that seeketh precise *truth*, had need to remember what every name he uses stands for; or else he will find himself entangled in words, as a bird in limetwiggs; the more he struggles, the more belimed. And therefore in Geometry, (which is the only Science that it hath pleased God hitherto to bestow on mankind), men begin at settling the significations of their words.
>
> THOMAS HOBBES

Introduction

Einstein's general theory of relativity has the reputation of being very difficult to understand. Indeed, it is often said that only six people in the world have mastered it. This is nonsense. The details of the theory are undoubtedly technical, but they can be understood by any mathematical undergraduate; indeed, examination questions are regularly set on them.[1] More to the point, its basic ideas can be explained using only simple mathematical concepts. We may count this as a piece of good fortune, since Einstein's theory has much to say about the possible structure and history of the universe. Indeed, the systematic study of cosmology dates from Einstein's application of his theory to the whole universe in 1917. There have been many developments since then, which bear on the problems raised by the uniqueness of the universe, that are discussed in

[1] Despite Einstein's belief that preparing for examinations tends to crush the research instinct.

the last part of this book. These problems can best be understood in terms of modern theories of cosmology, so to prepare the way we shall describe the general theory of relativity in this chapter, and its applications to cosmology in the next.

The Curvature of Space

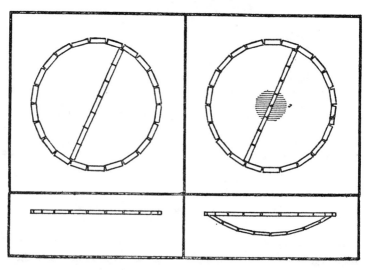

Fig. 47. (a) Measuring a circular disc which is at the same temperature throughout. The lower figure indicates that its cross section is flat. (b) The same disc when its center is cooler than its outer parts. The disc itself remains flat, but it appears to be curved because the rulers near its center have contracted and more of them are needed to complete a diameter. This apparent curvature signifies that the geometry of the disc as measured by the rulers is non-Euclidean.

The main feature of general relativity that goes beyond the theory of inertia described in Chapter IX is its contention that space is curved, or non-Euclidean. This idea has been the subject of much mystification, but it is actually very simple. To see how it arises, we must describe the way physicists interpret the measuring operations which they use to determine the properties of space. Let us begin with an analogy. Suppose a physicist is interested in a metal disc, which the makers claim to be circular. He decides to check this claim by measuring its circumference and radius, to see whether they are in the right relation—that is, whether their ratio is 2π, or 6.28. . . . He is careful to use a ruler so small that it can measure the circumference without sensible error. He is disturbed to find that, despite this precaution, the ratio he obtains is significantly less than the expected value. He is about to write an angry letter to the makers when he recalls that the disc was not in the standard condition specified by them. In fact the sun was shining unevenly on it, and its central portion, which was in a shadow, was colder than the edge (Fig. 47). The heat of the disc had been communicated to the ruler, and it had expanded. Moreover, this expansion was less when the ruler measured the radius than when it measured the circumference. The physicist realizes that this distortion of the ruler may have been responsible for his unexpected result, so he measures the temperature of the disc and makes the necessary corrections. He then finds that the makers are indeed right—the disc has a circular shape.

This illustrates the general principle that experimental physicists are not naïve; they do not take their measuring instruments for granted. But would it be so disastrous if they did? Suppose that our physicist had chosen to regard his ruler as the basic standard of length which never needs to be corrected. In that case, of course, he would find that the ratio of the circumference of the heated disc to its radius does not have the value to be expected according to Euclidean geometry. On the other hand, *the disc would still be a circle*, in the sense that every point on its edge would be at the *same* distance from its center.[2] In other words, the physicist would have dis-

---

[2] This assumes that the disc has been heated symmetrically, so that after its own expansion it is still circular. Alternatively, we could make the disc of invar, so that its expansion would be negligible.

covered by measurement that a disc which is a circle, in the sense that all its radii are equal, can have its circumference related to its radius in a non-Euclidean manner. He would then say that the geometry of the disc, as measured by his uncorrected ruler, is non-Euclidean (Fig. 47b).

Non-Euclidean geometries have been studied by mathematicians since the middle of the nineteenth century, so that their properties are now well understood. In particular, we know how to express mathematically the *amount* by which a geometry deviates from Euclidean geometry; in our present example this corresponds to the amount by which the ratio of the circumference to the radius of a circle differs from 2π. It is important to be able to express this deviation from Euclidean expectations in a numerical way, because we can then relate it to the distribution of temperature over the disc. In other words, we have a relation between the amount of deviation from Euclidean geometry and the physical state of the object concerned.

In practice a physicist prefers to avoid this situation—he always corrects his measuring instruments for thermal distortion. He does this, of course, partly in order to avoid having to learn non-Euclidean geometry. But there is also a more profound reason for his attitude. Suppose that he were to measure the disc with several rulers made of different materials. These rulers would expand by different amounts, and so each of them would register a different length. Now normally a physicist would correct his readings for all these distortions and so obtain consistent results. Suppose, however, that he adopted the other approach and made no corrections. He would then find that the amount of non-Euclidean geometry depends not only on the temperature distribution in the disc but also *on the ruler he uses*. In other words, the geometry of the disc would not be an intrinsic property of its thermal state but would depend on the ruler which is used to measure it. This is so inconvenient that no one would seriously suggest that the physicist should change his methods and work in terms of non-Euclidean geometry.

We now come to the main point of our analogy, which is that the claims of non-Euclidean geometry would be much stronger *if* all rulers expanded by the same amount. For then they would all represent a *consistent* standard of length whatever their temperature. Consequently the geometry of the disc, as measured by uncorrected rulers, would be *intrinsic* to the disc, depending only on its temperature distribution, and not at all on the material of the rulers. It might then be worth while adopting the non-Euclidean point of view as a convenient means of describing the state of the disc.

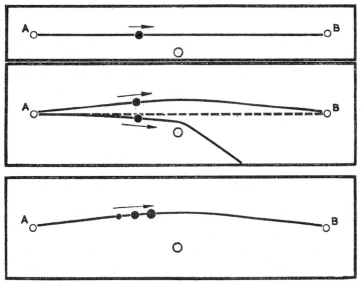

Fig. 48. (a) *The orbit of a moving body if gravitation were absent. The distance between A and B can be measured by the transit time.* (b) *The actual orbits when gravitation is present. The transit time from A to B is changed by the gravitational influence of the intermediate body.* (c) *Bodies of different mass move on the same orbit. Their transit times are all the same, so they all give the same value for the distance between A and B.*

This might seem a highly academic point, since we know that in fact different rulers do *not* expand by the same amount. However, the point is not at all academic when we come to measure the geometry of space. The main distorting agent is now gravitation rather than heat, and gravitation *does* distort all rulers equally. Since in practice only small distances are measured by rulers, we shall explain this property of gravitation in terms of rather more realistic measuring devices. We can determine the distance between two points by measuring the time taken by bodies of known speed to go from one point to the other. Light and radio waves are more often used in practice, but material bodies are better for purposes of exposition.

The nature of the gravitational distortion of our "rulers" is obvious: their motion is affected by gravitation and so is the time they take to go from one point to the other (Fig. 48b). One way of dealing with this situation would be to correct the measured time, so as to allow for the gravitational disturbance. The resulting distances between the points of space would then be related in the Euclidean manner.

On the other hand, we could choose to ignore the effect of gravitation on the bodies by making no correction at all; the geometry of space would then be non-Euclidean.[3] But we know from Galileo's experiment and the principle of equivalence that all bodies are *equally* affected by gravitation (Fig. 48c). They will thus all give the *same* uncorrected results for the distances between points. This means that the amount of non-Euclidean geometry will be the same *whatever body is used to explore the geometry of space.* In other words, the geometry of space is an intrinsic property of its gravitational state.

It was this reasoning that led Einstein to adopt the non-Euclidean point of view. Now, since space is said to be *curved* when it is non-Euclidean, we can say that Einstein's theory

[3] Strictly speaking, it is the geometry of *space-time* that would be non-Euclidean, but we do not need this refinement in our application of general relativity to the universe (Chapter XII).

THE GENERAL THEORY OF RELATIVITY

relates the curvature of space to gravitation. The name "curvature" is borrowed from the properties of spherical surfaces, whose geometry can be given a non-Euclidean interpretation. For instance, the shortest distance between two points on the surface of a sphere consists of the part of the great-circle that joins them (Fig. 49a).[4] Such a great-circle is thus the analogue of a straight line on a flat surface. Similarly, a triangle

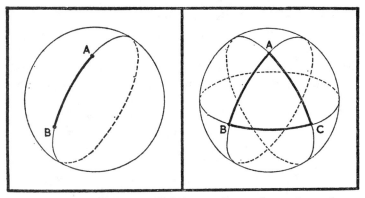

Fig. 49. A straight line and a triangle on the surface of a sphere. (a) The shortest distance between two points on the surface of a sphere is an arc of the great-circle joining them. It is the analogue of a straight line on a flat surface. (b) A spherical triangle.

formed of three great-circles is the analogue of a plane triangle (Fig. 49b). Now the sum of the angles of such a "spherical" triangle exceeds the value of 180 degrees that it would have in Euclidean geometry. Indeed, the amount of this excess is a measure of the deviation from Euclidean geometry, and is larger the smaller the sphere—that is, the more its surface deviates from a plane. Gauss discovered that the excess is just equal to the area of the triangle times the curvature[5] of the

---

[4] A great-circle on the surface of a sphere has the same radius as the sphere itself.

[5] Defined as the inverse square of the radius.

sphere's surface. This curvature, then, is a measure of the amount of deviation from Euclidean geometry.

We can easily understand what it means to say that a two-dimensional surface is curved, because we can see this surface lying in three-dimensional Euclidean space, and the meaning of the word "curvature" is quite obvious. But when this same name "curvature" is also given to a three-dimensional non-Euclidean space[6] it becomes rather misleading. The important thing to remember is that the curvature of the surface of a sphere is actually an *intrinsic* property of the surface. It does not depend on the fact that we can see the sphere lying in a three-dimensional Euclidean space. A two-dimensional insect, confined to the surface of a sphere, could detect the fact that the surface is curved by adding up the angles of a spherical triangle. In the same way the curvature of a three-dimensional space is an intrinsic property that can be detected by measurements in the space. All that is meant by the curvature of space, then, is that gravitation affects the motion of bodies, and that when this effect is ignored the geometry of space as measured by the bodies becomes non-Euclidean. What makes this point of view so attractive is that the amount of non-Euclidean geometry—that is, the curvature—does not depend on the bodies used to determine it.

Einstein's Field Equations

The next step is to express Einstein's idea in a quantitative way. In the case of the heated disc the curvature was determined by the temperature distribution and the expansion coefficient of the ruler. Correspondingly, we want to know how much curvature is produced by bodies like the sun, just as Newton's inverse square law tells us how much gravitation is produced by them. It was in order to answer this question that

[6] Let alone four-dimensional space-time!

THE GENERAL THEORY OF RELATIVITY

Einstein proposed his famous field equations. These equations, in fact, simply relate the curvature of space to the distribution of matter. In order to understand these equations one must know the mathematics of non-Euclidean geometry, but I hope it is clear how Einstein's equations arise and what sort of information they provide.

An important question is: How did Einstein decide which equations to choose? It turns out, for reasons that depend on the detailed mathematics of the problem, that the choice of equations was, in fact, very limited. On the basis of certain reasonable general principles he was led to a virtually unique set of equations. It was, therefore, a very striking result that the implications of these equations for the motion of planets around the sun not only reproduced those results of Newton's which in the nineteenth century were regarded as such a triumph and verification of Newtonian methods, but also succeeded in accounting for a discrepancy in the motion of Mercury.[7]

Einstein's equations also make predictions about the behavior of light in a gravitational field. The most famous of these predictions is that light is attracted by a massive body, so that its path would be bent out of a straight line. The direction of the source of light would then appear to be displaced. The effect is a small one, but it can be detected in the light coming from a star and just grazing the sun. Unfortunately such a star is visible only during the few moments of a total eclipse of the sun, so that accurate observations are difficult to make.[8] As a result, although the observed effect is roughly the right size, there is as yet no clear-cut numerical test of Einstein's prediction.

Until recently the same was true for another of Einstein's

---

[7] The theory also predicts a similar, but smaller, discrepancy for the earth. This prediction has recently been verified.

[8] Plate I shows one of the unsuccessful attempts to photograph stars at a total eclipse. All the stars were completely obscured by cloud.

predictions, namely, that light is reddened when it moves against gravity[9] (p. 136). For many years astronomers have attempted to detect this reddening in the spectrum of the light reaching us from the sun, and also from certain compact massive stars (white dwarfs) which have an exceptionally high value of gravity. Unfortunately, the Einstein red-shift is masked by astrophysical effects which distort the spectra, and despite many attempts its existence has not yet been established in this way.

These astronomical attempts have recently (1960) been superseded by laboratory experiments, which clearly demonstrate the change in the color of light[10] as it falls to the ground from the height of a few feet. These experiments are technically remarkable since the color change measured is 10^9 times smaller than that suffered by light reaching us from the sun. They were made possible by the discovery in 1958 of the so-called Mössbauer effect, which enables fractional changes in frequency of as little as one part in 10^{15} to be measured. This represents an accuracy many powers of ten greater than can be achieved in any other experiments.

It is clear from Einstein's success with the motion of Mercury and the behavior of light that his theory is more accurate than Newton's. But it will be remembered that Einstein's original aim in constructing his theory was not so much to achieve greater accuracy as to account in detail for the inertia of matter on the basis of Mach's principle. Even today there is considerable disagreement about the extent of his success in this respect. The main reason for this is the complexity of Einstein's theory—mathematical complexity, that is, because its physical foundations are quite clear.

What *has* been shown is that, if one body is brought near to another, then the inertia of each body will be increased a little, by an amount depending inversely on their distance

[9] This reddening can actually be deduced from the principle of equivalence without appealing to Einstein's field equations.
[10] In the form of γ rays.

apart. This is just the inverse first power law that was so crucial to our discussion in Chapter IX. At first sight this looks very promising for our program of accounting for the *total* amount of inertia in terms of the influence of all the bodies in the universe. Unfortunately, there arises at this point a difficulty which is peculiar to Einstein's theory. It arises because Einstein's theory has the property of being *non-linear*—that is to say, the combined gravitational force exerted by two bodies is not just the sum of the forces that each exerts in the absence of the other (Fig. 50). This means that we cannot now combine the influences of each of the stars as we did in the

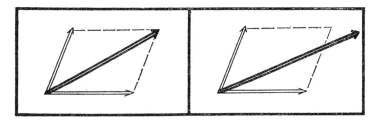

Fig. 50. Linear and non-linear addition of forces. (a) The usual method of combining forces. It corresponds to the vector law of addition. (b) "Non-linear" combination of forces. In this case, when two forces act simultaneously, their combined effect is not obtained by the vector law of addition.

simple theory, by just adding them together. The problem of combining them correctly leads to mathematical problems that have not yet been solved.

I must emphasize that this complicated non-linear character of Einstein's theory is not a defect. It is an essential part of Einstein's conception and, indeed, it is one of the most beautiful and important features of his theory. Unfortunately, it would take us too far away from the universe to pursue this here.[11] All we can say is that an intriguing problem awaits solution. Remarkably enough, the existence of this problem has not prevented Einstein's theory from being applied to the whole universe. This application, which underlies the last part of the book, is described in the next chapter.

[11] The non-linearity has to do with the fact that gravitational energy itself acts as a source of gravitation. This is actually required by Mach's principle!

CHAPTER XII

THE HISTORY OF THE UNIVERSE

> Ridiculous the waste sad time
> Stretching before and after.
> T. S. ELIOT

Introduction

The history of the earth is hard to fathom; to find out what happened long ago we must patiently infer the facts from fossils and the like. But the history of the universe is directly visible to us. As we look out to great distances we are seeing into the remote past. The use of light-time as the scale of distance is a constant reminder of this. When, for instance, Hubble showed that the Andromeda galaxy is a million light-years away, he was also showing that we see it now as it was a million years ago. At the limit of the 200-inch telescope we are seeing 7 billion years into the past.

Fantastically large though this time interval may seem, it is quite modest by cosmical standards. For instance, it takes about 14 billion years for a pair of receding galaxies to double their distance apart. It is true that we can see more than half this time into the past but, unfortunately, observations at the limit of the 200-inch telescope are not very accurate. We are just on the verge of seeing a substantial slice of the universe's past.

Until observations become more accurate, or stretch out to greater distances and remoter times, we have to turn to theory in order to discover the history of the universe. This is one of the many points at which cosmology becomes enveloped by controversy. For there exist several competing theories, each of which claims to be the best basis for extrapolating from our present limited knowledge to a complete temporal description of the universe. The "Establishment" view is that Einstein's general theory of relativity should be used, on the grounds that it is the best available theory of space, time, and gravitation, which of all the forces in nature appears to be the most important for the large-scale behavior of matter. Such a view commands respect, so this chapter gives a critical account of the relativistic theory of the universe. We shall need this account later when we come to discuss the uniqueness of the universe.

What we seek are those solutions of Einstein's equations which relate the geometry of the whole universe to the distribution of the matter it contains. Of course, in our first attempt to find such solutions we should not try to include *every* detail of the universe—we can restrict ourselves to its large-scale features. This procedure is reminiscent of that adopted in the study of the earth. There, too, large-scale features are first established. For instance, although mountains seem very impressive to us, they are only a small fraction of the size of the earth itself. Indeed, if the earth were scaled down to the size of a billiard ball its corrugations would become much smaller than those of actual billiard balls. This suggests that we can obtain a good understanding of the structure of the earth even if we suppose that its surface is smooth. Many geophysical phenomena can, in fact, be understood on this basis. Only later on, when we come to consider refinements, need we notice that the surface of the earth is actually rugged. In the same way we ought first to establish the main structural features of the universe. Finer details can come later (Chapter XIV).

The Cosmological Principle

The observations described in Part I strongly suggest that on a large scale the universe possesses a certain uniformity. One region large enough to contain many galaxies appears to be very like any other such region. In other words, when the universe is viewed on a large scale it lacks landmarks, like the surface of the earth when it is assumed to be completely smooth (no mountains, seashores, etc.). A sufficiently presbyopic man, placed blindfolded somewhere in the universe, would not, when the blindfold is removed, have any idea where he was. All he would see would be undifferentiated sameness.

This uniformity of the universe, if indeed it exists, is its most fundamental feature. The statement that the universe is, in fact, uniform is generally known as the "cosmological principle." This use of the word "principle" has led to a good deal of criticism, so I ought to emphasize that no *a priori* philosophical considerations are involved, as they are, for instance, in Mach's principle. It is not suggested that the universe "must" be uniform. It is a matter for observation whether the cosmological principle is true or false. Either possibility is quite consistent with the basic concepts of physics and astronomy, as they are understood today. On the other hand, if the universe is indeed uniform, as the evidence suggests, an explanation is called for. Consider the earth again: its approximate uniformity can be explained in terms of the gravitational action of its parts on one another, which pulls it into a uniform shape. The approximate uniformity of the universe also calls for an explanation in terms of the physical processes which govern its structure. This explanation has not yet been discovered.

While searching for this explanation cosmologists found that they could easily classify the many uniform structures which are consistent with our present observations. Using the language of general relativity, we can say that the uniformity of the universe implies that, at any given time, space has the same curvature everywhere. The purely spatial properties of the universe are thus entirely determined by just one number. Its present value can in principle be determined from observations of nearby galaxies. In practice, however, the observations so far made are not accurate enough to yield a reliable value.

Moreover, even if the observations were sufficiently accurate, we would not be satisfied. We really want to know far more about the universe than is summed up in the present value of its curvature. We want to know not only how the universe looks at different places but also how it behaved at different times—that is to say, we want to know the whole life history of the universe. What did it look like thousands of millions of years in the past, before the farthest galaxies we have yet detected emitted the light we now receive? What will it look like thousands of millions of years in the future when those galaxies will receive the light we now emit?

Since we are seeking simple over-all possibilities, our first choice must be that the universe does not change with time at all—that it always looks the same. This possibility is known as the steady state hypothesis, or the *perfect* cosmological principle. If it is correct the universe is uniform in time as well as in space, so that our unfortunate presbyope, in addition to being completely lost, would not know the time. No changing process going on around him would be available to act as a clock. This hypothesis, like the original cosmological principle, refers only to the *large-scale* structure of the universe of course; otherwise it would be easy enough to disprove it. Despite its extreme simplicity, the perfect cosmological principle played no part in the development of relativistic cosmology in the 1920s and '30s. It was not until Hermann Bondi, Thomas Gold, and Fred Hoyle advocated it in 1948 that it was recognized as a serious possibility. The reason for this remarkable delay will emerge from our account of what a *changing* universe is like.

Models of the Universe

At first sight it might seem that the universe could change in so many different ways that it would be very difficult to classify all the possibilities in an easily surveyable manner. Fortunately, if we adopt the cosmological principle by assuming that the universe is uniform in space, its change with time is easily described. This was discovered in 1936 by H. P. Robertson and A. G. Walker, who developed the pioneering work of the English cosmologist E. A. Milne (1896–1950) on this problem. They showed that, if the cosmological principle is satisfied, the history of the whole universe is determined by just one thing, namely the rate at which the distance between any pair of galaxies changes with time. The cosmological principle ensures that one obtains the same result whichever pair of galaxies one chooses.

This discovery enables us to classify very easily all the structures that satisfy the cosmological principle. First of all, at any given time the curvature of space must be the same everywhere but can have any value. Secondly, the distance between one pair of galaxies can change in any way with time, or be any "function" of time, as the mathematicians say. This function also determines the rate at which the curvature changes with time. Indeed, the geometrical properties of a particular structure are *completely* characterized by one constant (the curvature at any one time), and by one function of time (the separation between any one pair of galaxies). Each combination of a particular constant and a particular function corresponds to a *model* of the universe.

So far we have discussed only the geometrical properties of model universes. To find out how much *matter* there is in each model, we must use Einstein's field equations, which relate the curvature of space to its material contents. The fact that Einstein's equations are consistent with many possible models of the universe is one of the main weaknesses of our present theoretical approach. For there is, after all, only one actual universe. Since the various models consistent with Einstein's equations all have the same theoretical status, it is not clear why nature chose one of them to be the actual universe rather than any other. We can put the difficulty in another way. Until we formulate a scientific theory, we have placed no restrictions on how the universe can behave: as far as we can tell, anything might happen. The purpose of introducing a theory is to narrow down the possibilities to precisely those which are, in fact, realized in nature. A perfect theory will thus be like a total dictatorship, where everything that is not forbidden is compulsory. This complete rigidity is not achieved in Einstein's theory since, although it forbids some unrealized possibilities, it permits many others.

Unfortunately, no one has yet succeeded in finding a theory which is consistent with only one model of the universe. Cosmologists, therefore, proceed by trying to decide which of the relativistic models agrees best with observation. We might in this way discover the history of the universe before understanding its significance.

What, then, are the models like? Apart from one static model, in which galaxies remain a fixed distance apart, they all contain large-scale systematic motions of expansion or contraction. The static model and the contracting ones can be immediately ruled out since they are inconsistent with the redshifts in the spectra of the galaxies. Nevertheless, the static model is of historical interest, because it was the first one to be proposed—by Einstein, in 1917, in the famous paper that first related general relativity to the cosmological problem.

At that time the red-shift was not well established, so that a static model might seem to have been a reasonable possibility. However, as we saw from the discussion of Olbers' paradox, in such a universe the amount of light in the night sky would be the same as that at the surface of a star. Had Einstein remembered, or reproduced, Olbers' argument, he would have seen immediately that his static universe was in disagreement with observation in the most violent way. He would then have been forced to propose an expanding model.

The first such model was actually discovered in the same year (1917) by the Dutch astronomer Wilhelm de Sitter. In the twenties other expanding models were discovered by the Russian A. Friedmann and the Belgian Abbé Georges Lemaître. By a remarkable coincidence it was during these same years that the expansion of the universe was established observationally.

These various models can be classified in physical terms as well as geometrically. Consider, by way of analogy, the behavior of a ball thrown up from the earth. If the ball is thrown slowly it will fall back to earth and then bounce up and down indefinitely (neglecting air resistance and the energy loss at each bounce). If we steadily increase the velocity with which the ball is thrown up we shall eventually reach a velocity at which the ball just escapes from the earth and no longer returns. This limiting velocity, the so-called escape velocity, which is of great importance in the design of rockets intended to reach the moon, is actually about 7 miles per second. If the ball is thrown faster than this it gets away from the earth with energy to spare.

The ball, then, can behave in two essentially different ways, according to its velocity. In the first, the ball is gravitationally bound to the earth and is able to get only a finite distance away. It then moves in a cyclic fashion up and down. In the second, the ball moves indefinitely away from the earth.

These two types of behavior are exhibited by the various non-static models of the universe. The similarity springs from the fact that, as the universe expands, the gravitational attraction of the galaxies tends to slow the expansion down. If the expansion is slow enough gravitation is able to stop it completely and to turn the expansion into a contraction. The universe then collapses onto itself until its density is very high. It will then re-expand at the same rate as before, and so on. It thus behaves in a cyclic fashion, just like the bouncing ball. On the other hand, if the velocity of expansion is great enough gravitation never succeeds in turning expansion into contraction, and the universe is able to expand indefinitely.

All these models have a singular moment in time, when the density of matter in the universe is infinite. In the cyclic model, indeed, there are an infinite number of such moments. In the other models there is just one, which is often loosely referred to as the creation of the universe. This is a very misleading phrase, since no *process* of creation is implied by the theory; if one goes backward in time from the present one simply finds that the models do not contain moments earlier than the time when the density was infinite. Presumably the theory ceases to apply even before this, when the density is so high that the atomic structure of matter begins to play an important part.

The Steady State Model

We are now in a position to appreciate why the unchanging universe of the perfect cosmological principle was not adopted at the outset. Although this principle is satisfied in Einstein's static model, that model was ruled out by the discovery of the expansion of the universe. It was naturally assumed that as a result of the expansion the density of matter in the universe is growing less and less. This would mean that the universe *is* changing with time. Our presbyope could then use the universe itself as a clock: he could tell the time simply by measuring the density of matter.

Nevertheless, the perfect cosmological principle has a seductive simplicity about it. Bondi, Gold, and Hoyle felt that this simplicity was so compelling that it ought to be preserved in the face of anything except *direct* observational evidence. That is to say, no extrapolation from observation, however familiar it might be, was to count against the principle—the only thing that counts is the observations themselves. They were then able to reconcile the principle with the expansion by suggesting that new matter is being continually created so as to maintain a constant density. This suggestion is, of course, in-

THE HISTORY OF THE UNIVERSE

consistent with the conservation of matter, but since the required creation rate turns out to be only about one hydrogen atom in a liter every trillion years, no observation is contradicted, but only an extreme extrapolation therefrom.

Since 1948 there has been much discussion, both scientific and philosophical, of the idea of continual creation. It is difficult to summarize this discussion, but perhaps its most interesting feature is that astronomers seem more ready than nuclear physicists[1] to accept the possibility that matter can be created. It is easy to make sacrifices on other people's behalf! My own attitude to the idea of continual creation will emerge in later chapters, where some of its consequences are explored. But a few words should be said here. First of all, there is no known observation which is in conflict with the idea. This is important since, although the postulated rate of creation is too small for the creation process itself to be directly observed, the steady state hypothesis has many astronomical consequences by which it can be judged. It is possible that researches now in progress will be able to decide between the cosmological principle and the perfect cosmological principle, but at the moment this decision cannot be made.[2]

Secondly, the idea of continual creation increases the range of phenomena that can be studied scientifically. Instead of accepting the universe as given, it regards it as the result of a set of *processes* which can be the subject of scientific investigation. A typical problem would be a study of the factors which determine the rate of creation: How is it affected by nearby concentrations of matter? Is it mainly governed by long-range forces (p. 129)? Does it ensure that the large-scale structure of the universe is uniform? And so on.

---

[1] Although since the breakdown of reflection invariance has been discovered their faith in the absolute validity of conservation laws has been noticeably weakened.

[2] I have given an account of the present (mid-1958) observational status of the problem in an article in *Vistas in Astronomy*, Vol. 3. It is somewhat more technical than this book, but not very much so.

It should be emphasized that this increase in range is a *provisional* advantage, in the sense that one is influenced by it only when the available observations do not yet distinguish between the various possibilities. Such arguments have their place in science, because the time available for research is limited, and they suggest which theories are likely to be the most fruitful to work on. Nevertheless, the validity of the steady state model can be decided only by observation. Moreover, if it turns out to be correct it can be understood only when we possess a theory which leads to it uniquely. We may hope that the observational decision will be made in the near future, but the theoretical one will, I fear, take much longer.

CHAPTER XIII

THE UNIQUENESS OF THE UNIVERSE

> and it is great
> To do that thing that ends all other deeds,
> Which shackles accidents, and bolts up change.
> SHAKESPEARE, *Antony and Cleopatra*

Introduction

In the introduction to Part II of this book great stress was laid on the fact that the universe is unique. Since the universe is, by definition, the totality of things, its uniqueness is self-evident. Nevertheless, it was emphasized because when one sets out to study a unique system one soon runs into difficulties of a kind not normally encountered in scientific work. This is not, perhaps, immediately obvious. As Bertrand Russell said in another connection, "The first problem is to see that there is a problem." We have had a hint of the difficulties already; we saw earlier how unsatisfactory it is that general relativity is consistent with many models of the universe instead of with just one. My aim in the remainder of the book is to explore these difficulties in detail. They will be stated in this chapter, and an attempt will be made to resolve some of them in the last two chapters.

Laws of Nature and Initial Conditions

The fundamental point is that a theory designed to describe a unique system should not contain any arbitrary features. To explain what is meant by this, it is best to go back to Galileo and Newton, and to examine various features of their scientific method in action. How did they approach a particular problem—for instance, the motion of projectiles? Their basic idea was to compare many examples of such motion, in the hope

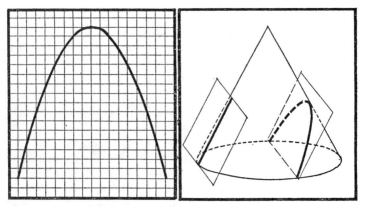

Fig. 51. A parabola. It can be obtained by cutting a cone with a plane parallel to a tangent plane.

of finding features common to them all. They soon succeeded; at whatever angle a projectile is thrown, and with whatever speed,[1] its orbit is always a parabola (Fig. 51).

They went on to explain this result in terms of the action of a gravitational force, but this development is only an extension of the basic idea, which is to find a common element in different situations. This common element then represents a property of nature which is beyond our control. We can choose

---

[1] Within certain limits.

the height and length of the orbit by the way we throw the projectile, but we cannot prevent its moving in a parabola.

This idea is now so familiar that it is, perhaps, hard to realize just how enlightening it must have seemed in the 17th century. The scientists of that time, faced with the problem of analyzing a great mass of facts, now found their task much simplified. They classified these facts into sets with common features, and looked upon these features as fundamental and unalterable properties of the world, which they dignified with the name "laws of nature." It is this ability to extract a common pattern, a law of nature, from a mass of facts which has been mainly responsible for the success of science since Newton's time, a success that was summed up in Pope's famous couplet:

> *Nature and Nature's laws lay hid in night:*
> *God said, Let Newton be! and all was light.*

It is clear that an essential feature of this method of analysis is the existence of several comparable situations. By examining them all we try to discover which are the fundamental features, like the parabola, which can then be elevated into laws of nature, and which are, as it were, the accidental features, like the initial speed of any particular projectile. If we knew of only one instance of a phenomenon we would not be able to tell which of its features were fundamental and which accidental. For instance, if we were able to observe only one example of a projectile being thrown, then for all we could tell, the fact that it reaches a particular height might be just as important as the fact that it moves in a parabola.

One way around this difficulty was discovered by Newton when he was studying the motions of the planets around the sun. This was the only system of its kind available, so the question arose: Is the fact that all the planets move in ellipses a more fundamental property of the solar system than the fact that their distances from the sun are as observed? Newton believed that the elliptic motion is indeed more fundamental;

that the planets might have been at other distances, but that they would then still move in ellipses. Newton was able to justify this belief, despite his knowing only one solar system, by means of a brilliant device. He assumed that the motion of the planets under the gravitational attraction of the sun could be compared with the motion of projectiles under the gravitational attraction of the earth. The solar system then ceases to be unique. The feature common to all these motions is the inverse square law of gravitation, which, in particular, implies that the planets must move in ellipses whatever their distances. Newton accordingly concluded that the ellipses are fundamental and that the distances of the planets are not.

This was at the time a tremendous step forward. Nevertheless, the idea that certain features of the universe are accidental is a somewhat disturbing one. It is possible that they would not appear accidental in a more comprehensive theory. This doubt is strengthened by the existence of evidence that the actual distances of the planets from the sun may not be at all accidental. For these distances satisfy a relation discovered by Titius and then taken up and widely publicized by Bode (1747-1826). According to Titius and Bode, the average distances of the planets from the sun are related in a simple way to the successive powers of 2, the distance to the nth planet, taking the earth's distance as 1, being $0.4 + 0.3 \times 2^{n-2}$, except for Mercury ($n = 1$), whose distance is 0.4.

The accuracy of this relation is shown by the following table.

PLANET	BODE'S RELATION	OBSERVED
Mercury	0.4	0.39
Venus	0.7	0.72
Earth	1.0	1.00
Mars	1.6	1.52
—	2.8	—
Jupiter	5.2	5.20
Saturn	10.0	9.54
Uranus	19.6	19.18
Neptune	38.8	30.07
Pluto	77.2	39.46

The agreement is very good except for Neptune and particularly Pluto. This discrepancy may not be significant, since it has been suggested by the English mathematician and astronomer R. A. Lyttleton that Pluto was once a satellite of Neptune which escaped under the influence of another satellite. It may be objected that Bode was only being wise after the event, since any given set of distances could be fitted by some relation. However, Bode's relation passed the same test as Newton's inverse square law: it made correct predictions. For Uranus was not discovered until 1781, *after* Bode's relation had been published. Furthermore, the gap in the planets at a distance of 2.8 was later shown to be filled by several minor planets, which some believe to be the debris of a large planet that exploded. These successes led people to call the relation Bode's *law*.

It is significant that in recent years at least one theory of the origin of the solar system has succeeded in accounting for Bode's law. This suggests that what for Newton was an accidental feature of gravitational phenomena may actually be just as fundamental as the law of gravitation itself. It appears, then, that, as more considerations are taken into account, fewer features of the universe remain accidental. Inevitably we are led to ask the question, Does this process ever stop? The law of gravitation taken by itself implies a distinction between the fundamental and accidental features of a phenomenon, but this distinction gets less clear-cut as we take more things into account. It looks as though by the time *everything* is taken into account the distinction implied by the law is—a legal fiction!

Now when everything is taken into account one is involved with the whole universe. We may suspect that there is then nothing accidental left. Unfortunately, this suspicion is not immediately confirmed. The reason for this is that the methods used to remove accidents have a serious limitation. For example, the distances derived by the theory underlying Bode's law are not absolute; they depend on the mass of the original star out of which the solar system was formed. This mass is now the accidental feature, and will remain so until a theory of star formation is established. This theory, in turn, will start out from an accident, namely the density of matter in the clouds of gas from which the stars have presumably formed. And so it will go on. At each stage one starts off the calculation with an accident—an *initial condition*, as it is called. Only when this initial condition is known or assumed can the behavior of the system be determined from the laws of nature.

This means that by themselves the laws of nature do not determine the behavior of the universe. Indeed, they were specifically devised to involve an *arbitrary* initial condition; this gives them just the flexibility they need to account for the variety of local phenomena. But this flexibility becomes an embarrassment when the laws are applied to the single system that is the whole universe. To avoid this embarrassment, we must find some way of eliminating the need for an initial condition to be specified. Only then will the universe be subject to the rule of theory.

The Contents of the Universe

There is another way of putting the problem. An atomic physicist studies what one might call the *typical* behavior of matter. He will tell you, for instance, how a hydrogen atom behaves in any given environment. But in his professional capacity he is not interested in the environments that hydrogen atoms actually have in the universe, or how many there are, or any other *actual* fact. All his methods and techniques are designed to study just typical behavior. On the other hand, the cosmologist is interested in what actually is the case; he is the naturalist of physics. The fact that most of the atoms in the universe are hydrogen, while only a few are heavier elements, has to be understood by him just as much as the way in which each atom behaves. Again, the fact that some of these atoms are actually concentrated in relatively small regions, namely galaxies, which have a well-defined average mass and separation, must be understood too.

What the cosmologist requires, therefore, is a theory which is able to account in detail for the contents of the universe. To do this completely it should imply that the universe contains no accidental features whatsoever. This provides us with a criterion for assessing the validity of rival theories. We believe this criterion to be so compelling that the theory of the universe which best conforms to it is almost certain to be right. In the next two chapters we use the criterion to compare the various models of the universe which are consistent with the general theory of relativity. Of course, the fact that general relativity leads to many possible models implies that that theory itself is inconsistent with our criterion. Nevertheless, one of its models stands out, because it is the only one in which the properties of galaxies and the cosmical abundances of the elements can be calculated without any accidental initial conditions. This model is the one which satisfies the perfect cosmological principle—that is, the one in which the continual creation of matter maintains the universe in a steady state despite its expansion. According to our criterion, a theory more comprehensive than general relativity would lead to this model uniquely. It will be interesting to see whether this conclusion turns out to be correct.

CHAPTER XIV

THE FORMATION OF GALAXIES

> Do I dare disturb the universe?
> T. S. ELIOT

Introduction

The models of the universe that were described in Chapter XII are all exactly uniform. They ignore the concentration of matter into galaxies, and thereby represent an approximation to the actual universe analogous to ignoring mountains and regarding the earth's surface as perfectly smooth. Such models are very useful, for they provide an over-all picture of a complicated system. Nevertheless, it must be remembered that they are only an approximation. The deviations from exact uniformity are actually of great interest, both in themselves and also as clues to the past history of the system. These deviations have consequently become the center of attention, and the way in which both the mountains and the galaxies were formed is now being hotly disputed by geophysicists and cosmologists.

One difficulty in both problems arises from the fact that the mountains and galaxies we now see were formed a long time ago. We must first discover, then, what the earth and the universe were like at the relevant times. As regards the earth, some geophysicists think that it was very hot when it was formed, and that it has been cooling down ever since. Others think that it was formed cold and that it has been heating up as a result of internal radioactivity. Until this dispute is settled, the best procedure is to investigate how the mountains might have been formed in each case. If one set of mountains looked more like the real thing than the other, we would probably have discovered not only how the mountains were formed but also which version of the earth's origin is correct.

For the same reason we must consider how galaxies might have been formed for each possible history of the universe. Of the various possibilities we shall distinguish only two: evolving models, in which the universe was much denser in the past than it is now; and the steady state model, in which the universe was the same in the past as it is now. Various further subdivisions can be made, but they involve a change of detail only, not of general principles.

Before these two possibilities can be discussed in detail some basic assumptions must be made. First of all, we must decide what a galaxy is for our purposes—that is, which of its properties can be taken as its defining characteristic, in terms of which its other properties can be subsequently derived. The general consensus is that the *basic* property of a galaxy is neither its stars nor its spiral arms but simply the large concentration of matter inside it as compared with the concentration in between the galaxies. This difference is considerable; inside a galaxy there is about one hydrogen atom per cubic centimeter, whereas in intergalactic space there is only about one hydrogen atom in a hundred liters, which corresponds to a ratio of 100,000:1. What has to be explained, then, is the existence in the universe of localized concentrations of matter.

Secondly, we must make some assumption about how a galaxy holds itself together against whatever influences are causing the expansion of the universe. This local lack of expansion can, in fact, be explained by the large density that is characteristic of a galaxy—the resulting gravitational attraction is strong enough to hold it together. Actually a galaxy is not the largest structure which ignores the general expansion. The record seems to be held by clusters of galaxies, which recede from one another but do not themselves expand. Presumably their gravitational attraction is strong enough to hold each cluster together but too weak to bind one cluster to another. The existence of clusters must therefore be explained as well as the existence of galaxies.

In practice the gravitational attraction of a galaxy has more to do than just overcome expansion; it has to hold the galaxy together against the dispersive influence of heat. This is an important function of gravitation, because before any stars are formed a galaxy is entirely gaseous. This gas may be very hot, which would mean that its atoms are moving about randomly and at high speed. The galaxy would then disperse unless it were held together by gravitation. In fact the gravitational forces must be so strong that the escape velocity from the galaxy exceeds the velocities of most of the atoms.

In order, then, for a galaxy to maintain itself as a distinct non-expanding unit in an expanding universe, gravitation must be the dominant force in a localized region of space, but not in the universe as a whole. Can the existence of such deviations from uniformity be explained in terms of models which are themselves uniform? Can these models be disturbed? The answer is that they can. Since the disturbances arise in characteristically different ways in the evolving and steady state models, we shall consider these models separately.

Evolving Models

According to the evolving models, the universe used to be very much denser than it is now. Indeed, about 10 billion years ago the density was theoretically infinite. At those early times the universe consisted of very hot gas; so hot, in fact, that gravitation had no chance to dominate, even in localized regions. However, the gas cooled down as it expanded, and after about a million years its heat was no longer able by itself to prevent galaxy formation.

This does not necessarily mean that the gas could then break up into galaxies. For the universe was still expanding, and this expansion would have had to be overcome in localized regions before galaxies could form. How was the expansion overcome? The basic idea is that, if for some reason a region of space temporarily contains far more than the normal number of atoms, then the gravitational effect of these atoms on one another has some chance of holding the concentration together.

This idea, which has been studied by many theorists, is described in non-technical language in George Gamow's book, *The Creation of the Universe* (1952). As Gamow explains, the reason why certain regions temporarily contain more atoms than normal is that, in addition to its general motion of expansion, the gas is swirling about in a random way; to use a technical phrase, its motion is highly turbulent. As a result of this random turbulent motion the density of the gas will not be uniform; it will be greater than normal in some regions and less than normal in others. Each localized concentration that is formed in this way will neither be swirled away again nor will it re-expand, if its gravitational attraction is sufficiently strong. We shall then be left with some permanent regions of higher concentration than normal, which no longer share in the expansion of the universe but which remain distinct and distinguishable objects. These objects are identified as clusters of galaxies.

What decides whether the gravitational attraction is sufficiently strong? In other words, what decides whether most of the atoms have less than the escape velocity from the region of increased concentration? In the first place, the larger such a region is the stronger is its gravitational attraction, and the larger is the escape velocity from it. In contrast, the actual velocity of the atoms will be the same whatever the size of the region. Thus the larger the region the more chance gravitation has to hold it together.

Nevertheless, we cannot contemplate indefinitely large regions as potential galaxies. The character of the turbulence sets a limit to the size of the largest regions in which a concentration can occur. If we were to examine a larger region we would find a density increase in some parts of it and a density decrease in others, which would tend to balance out. Very rarely would we find an appreciable increase in the whole region. The size of a galaxy can thus be calculated.[1] It cannot be too small, or turbulence would swirl it apart, and it cannot be too big, or turbulence would not have brought it together.

It is at this point that we meet the difficulty we discussed in the last chapter. The general theory of relativity does not determine the character of the turbulence which prevailed when the universe was a million years old. If the theory we are describing is correct, the observed sizes of galaxies can be used to tell us this character. From this we could calculate what the turbulence was like at the beginning of the expansion. But no reason is given why the turbulence had just these characteristics, rather than quite different ones. Its actual characteristics are thus an accidental initial condition devoid of theoretical significance. As a result the sizes of galaxies are also purely accidental.

This state of affairs, while logically possible, is very unsatisfying. We should surely keep to a minimum those elements of our experience which we are forced to regard as arbitrary. Fortunately, in the second type of model—the steady state model—the properties of galaxies are not at all accidental. Let us see how this comes about.

Steady State Model

Our approach to the problem of galaxy formation is fundamentally different when we adopt the steady state model. For there is now no question of there being a special moment of time which was favorable for galaxy formation. If the universe is expanding, but in a steady state, new galaxies must be forming regularly at all times to replace those that recede. We must therefore show that conditions in the universe are *always* favorable for the formation of galaxies.

It should be emphasized that our assumption that the uni-

[1] In practice the calculation meets severe technical difficulties which have not yet been overcome.

verse is the same at all times does not in itself guarantee that galaxies can form. It might turn out that conditions are *never* favorable for the formation of galaxies. Such a possibility is consistent with the universe being in a steady state, but of course this model would then be in violent disagreement with observation. Fortunately this difficulty does not arise, and for a reason that is peculiar to a system in a steady state. In contrast to the evolving models, there is here no such thing as the *first* set of galaxies, forming in a universe that has not hitherto contained any. We do not have to *create* a population of galaxies, but only to maintain it.

This distinction is of the greatest importance. For now all we need show is that galaxies form a *self-propagating* population.[2] Self-propagation does, in fact, occur because a given set of galaxies exerts gravitational forces on the gas that lies between them, producing in it large density concentrations. These concentrations will then develop into new galaxies, which form the next generation. Moreover, this self-propagation removes the arbitrariness in the properties of galaxies. The reason for this is that, in a steady state, the child galaxies must have the same properties as their parents. *This requirement suffices to determine these properties completely.* As a first step to understanding this, let us see in more detail how one galaxy can give birth to another.

We fix our attention on a particular galaxy which is going to act as the parent in a birth process. Since our aim is to calculate the properties of galaxies, we do not yet know anything about this galaxy. Nevertheless, for reasons which will emerge, we assume that it is moving through the intergalactic gas in its neighborhood. We want to study how this gas is affected by the galaxy's gravitation, so it is more convenient to suppose that the galaxy is at rest and that the gas is streaming past it. This gas is attracted by the galaxy, and so moves in the manner shown in Fig. 52. As a result of this motion the gas

---

[2] Strictly speaking, we must also show that this self-propagating population is stable, and that a complete absence of galaxies is unstable. This can be done.

in the wake of the parent galaxy is compressed, and in the shaded region of the figure its density will be large enough for it to collapse under its own gravitation.

We have here a mechanism for producing density concentrations that is much more systematic in its action than the random turbulent motion postulated in the evolving models.

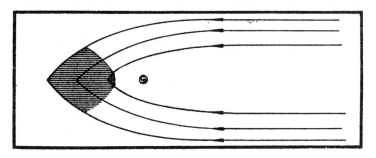

Fig. 52. The birth of a galaxy. Intergalactic gas moves past the parent galaxy and is attracted toward it. The gas in the shaded region has a large enough density for it to collapse under its own gravitation, thus forming a new galaxy.

As a result it is possible to calculate the conditions under which a concentration can collapse under its own gravitation, without having to solve the technical problems associated with highly turbulent motion. This calculation is based on the following facts:

(i) The size and mass of the region that can collapse depend on the temperature of the gas—that is, on the random motions of its atoms. These motions must be less than the escape velocity from the region.

(ii) The temperature of the gas depends on its degree of compression. (All gases heat up when compressed.)

(iii) The degree of compression depends on the gravitational attraction of the parent galaxy, and so on its mass (Fig. 53).

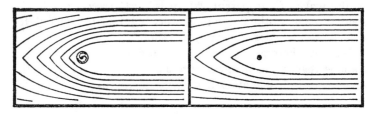

Fig. 53. The effect of galaxies of different mass on the intergalactic gas. The smaller the mass of the parent galaxy the less the intergalactic gas is compressed.

We thus have the following chain of relations:

| Mass of→ parent galaxy | degree of→ compression | magnitude→ of random motion | mass which can control random motion |

Through this chain of relations the mass of the child galaxy is determined by the mass of its parent.

Now if the system of galaxies is to be in a steady state, parent and child must have the same mass. However, according to the calculation we have just outlined, they will not have the same mass unless the parent weighs between about 10^{43} and 10^{44} grams. If a particular galaxy were to have a much greater mass, say 10^{50} grams, then its child would be lighter, and its child lighter still . . . until a mass around 10^{44} grams is eventually reached. Subsequent generations would then all have about this mass. On the other hand, if a galaxy were to be much lighter than 10^{43} grams, its progeny would be heavier and heavier, until a mass of 10^{43} grams is reached. Subsequent generations would now all have this mass. Thus if the universe is in a steady state we should expect to find the masses of most galaxies lying in the range 10^{43}–10^{44} grams. The observed masses of galaxies do indeed lie in this range.

We have here the first example of an actual property of the universe being calculated from general principles, without the intervention of any arbitrary initial conditions. This calls for comment, but we shall first follow up the fate of a child galaxy, since this sheds more light on the way the steady state condition restricts the properties of galaxies.

The Formation of Clusters

What happens to a child galaxy depends on how fast it is moving away from its parent, which, in turn, depends on the velocity of the original gas. If the child has less than the escape velocity from its parent, the two galaxies are bound together and form a double galaxy. On the other hand, if the child has more than the escape velocity, it will succeed in getting completely away from its parent, and another single or "field" galaxy will have been formed.

We can now contemplate another birth process, since the supply of intergalactic gas has been replenished by the creation of new matter. If the child succeeded in escaping its parent the new birth process will be just like the old, each field galaxy acting independently as a parent. A new situation only arises if, in the first birth process, a double galaxy was formed.

Such a double galaxy acts as a parent in the same general way as a field galaxy. One important difference is that it exerts a larger gravitational influence than does a field galaxy, and so compresses enough material to make two new galaxies. Whether this material in fact forms one large or two normal galaxies depends on the details of each case. Another consequence of the larger mass of the parent is that the children are now almost certain to be captured. A triple or quadruple cluster will thus be formed.

This cluster, in its turn, acts as a parent. It will compress enough material to make three or four new galaxies, which will all be captured. With each generation, then, the cluster more or less doubles its size, so we have here a large cluster in the making.

What does the whole system of clusters look like at any particular time? The size of a cluster is clearly determined by its age, so what we want to know is the distribution of ages. This can easily be calculated.[3] One of its main features is that in any particular region of space most clusters are younger than five billion years. The reason for this is that very old clusters must be very far apart, since they have been receding from one another for a very long time. Thus, even though the steady state universe is infinitely old, most clusters in any given region are no older than the Milky Way (whose age is estimated to be about five billion years). Corresponding to this comparative youth, the average number of galaxies in a cluster is only about ten. In addition there will be about a hundred large clusters (of a thousand or more galaxies) within a hundred million light-years.

This theoretical picture is nowhere in conflict with observation, but in describing it we have overlooked one point which threatens the whole edifice. This point shows how exacting is the requirement that the system of galaxies be in a steady state. For consider how many birth processes are needed during the time that the galaxies move so far apart as to occupy twice their original volume. In that time the number of galaxies must be doubled, if their number per unit volume is to be maintained at a fixed value. Now I have just outlined how this doubling might take place. But as it stands this mechanism has a fatal flaw—each original field galaxy does not produce one new field galaxy, since some of them capture their children to become double galaxies. This means that at the end of the birth process there will be fewer field galaxies per unit volume than there were at the beginning.

As time goes on the number of field galaxies gets less and less, until eventually there are hardly any left. But in consequence of this there will be fewer double galaxies forming, since they are formed from field galaxies. Similarly, if there are fewer doubles, then in later generations there will be fewer

---

[3] The age distribution is exponential, the average age being one third of Hubble's constant.

triples, quadruples, and so on. In other words, by the time a steady state is reached there will be no galaxies at all!

There is only one way of avoiding this unfortunate state of affairs. There must be some other source of field galaxies which we have overlooked. Fortunately there is such a source. The galaxies in a cluster are constantly exchanging energy with one another, as a result of their mutual gravitational influence. Occasionally a galaxy will gain so much energy that it will be able to escape from the cluster. This process is analogous to that of evaporation from a liquid surface, where an atom or molecule occasionally gets enough energy to overcome the attractive forces of the other atoms, and leaps away from the liquid. Here, then, in the evaporation of galaxies from clusters, is a new source of field galaxies. Incidentally, as a result of their high velocity, these field galaxies will be moving through the intergalactic gas—it is this motion which we previously invoked without explanation in our first description of a birth process (p. 174).

Although this is a plausible source of field galaxies, we are still faced with a serious problem. For if the system of galaxies is in a steady state, the number of field galaxies produced by evaporation must just equal the number lost by capture. But there is no *a priori* reason why these two numbers should be equal; evaporation might be a very slow process, for instance. Fortunately it turns out that if at any time these numbers are not equal the contents of clusters will gradually change until they are. The resulting system will then be in a steady state. It is precisely this system that we have been describing.

Conclusions

We thus see that, as a result of some rather intricately linked processes, galaxies will exist in the steady state model of the universe, and that the properties of these galaxies are in general agreement with observation. This agreement should not be pressed very far at the moment—both the calculations and the observations are rather inaccurate. What should be emphasized is that all the numerical results can be calculated from the laws of motion and of gravitation, and fundamental constants like the velocity of light and the gravitational constant. No arbitrary initial conditions are involved.

How has this result been obtained? What has been added to the laws of nature to make them determine the contents of the universe to this extent? The answer is the hypothesis that the universe is in a steady state. For only one set of galactic properties leads to an accurately self-propagating system; this set of properties must then be realized in nature if the hypothesis is correct. The steady state condition here replaces the initial conditions normally required, but now no arbitrary numerical quantities have to be specified. In the steady state model the properties of galaxies are as intrinsic to the universe as the laws of nature themselves. We regard this as a compelling argument in its favor.

CHAPTER XV

THE FORMATION OF THE ELEMENTS

SUBTLE:
                       Ay, for 'twere absurd
To think that nature in the earth bred gold
Perfect in the instant: something went before.
There must be remote matter. . . .
Nor can this remote matter suddenly
Progress so from extreme unto extreme,
As to grow gold, and leap o'er all the means.
Nature doth first beget the imperfect, then
Proceeds she to the perfect.
           BEN JONSON, *The Alchemist*

Introduction

With the simple substitution of iron for gold, Subtle's remarks would form an admirable introduction to a modern account of the formation of the elements. In this chapter we shall describe the relation between the various model universes and the way nature proceeds from the imperfect to the perfect. We shall also see why iron is the perfect element.

The Structure of the Atom

We begin by summarizing the relevant features of atomic structure. The various elements can be arranged in a series of increasing weight, beginning with hydrogen and ending with uranium.[1] By convention, oxygen is given the atomic weight 16. On this scale the atomic weight of hydrogen is very slightly greater than 1 (1.008), and that of uranium is slightly greater than 238 (238.07). Most elements have an atomic weight that is nearly a whole number.

This appearance of whole numbers served as a valuable clue in the elucidation of atomic structure. A further clue was later found to be contained in the small deviations of atomic weights from exact whole numbers. But before these facts were fully understood an explanation had been found for another important feature of the elements, namely that their chemical properties vary with their atomic weight in a roughly periodic way.

This fact led the Russian chemist D. I. Mendeleef (1834–1907) and others to construct the famous periodic table of the elements, in which they are placed in order of increasing atomic weight, in such a way as to emphasize the regularity in their properties. The chemical behavior of a particular element is largely determined by its position in this table: all alkali metals occupy one vertical column, all inert gases another, and so on. This periodicity in the chemical properties of the elements remained a mystery until the discovery in 1911–13 of the Rutherford-Bohr model of the atom (Fig. 54).

An atom consists of a central positively charged nucleus, which contains most of its weight, around which negatively charged electrons are circulating. This structure is often compared with the solar system, but there are important differences. In the first place, the nucleus controls the electrons by electrical forces rather than the much weaker gravitational ones. Secondly, and perhaps more importantly, there is an essential difference in the way the electrons and the planets are arranged in space. The electrons do not obey Bode's law, or anything like it. Instead, they are arranged in shells (Fig. 54).

These shells are important because the chemical properties of an atom are mainly determined by the number of electrons

[1] Transuranic elements have recently been formed artificially.

outside the largest complete shell. For instance, alkali metals all have one outer electron, and inert gases have none. Since the number of outer electrons changes periodically with increasing atomic weight, the chemical properties will also change periodically.

This explanation of the periodic table was a great achievement for atomic theory, but by no means an isolated one. In

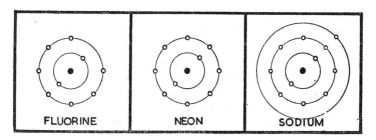

Fig. 54. *The Rutherford-Bohr model of the atom. Negatively charged electrons circulate around a heavy positively charged nucleus. These electrons are arranged in shells; the number of electrons outside the largest completed shell determines the most important chemical properties of the atom. The diagram shows a halogen, an inert gas, and an alkali metal, for which the outer shell lacks one electron, is complete, and has an extra electron, respectively.*

a few years it succeeded in accounting for a vast range of phenomena, and, in particular, for the spectra of the elements. The detailed structure of the nucleus played little or no part in this success. However, it *was* needed to explain an important phenomenon established by the English physicist J. J. Thomson in 1913, namely the existence of substances with the same chemical properties but different atomic weights. Such substances occupy the same place in the periodic table and so are called isotopes.

To explain the existence of isotopes, we must give a brief

account of nuclear structure. The nucleus contains two types of particle, protons and neutrons. A proton is about 2000 times heavier than an electron, and carries a positive charge numerically equal to the negative charge on an electron. A neutron is slightly heavier than a proton and carries no charge. Since an atom is electrically neutral, its nucleus contains as many protons as there are electrons circulating around it; since protons are so much heavier than electrons, this explains why most of the mass of an atom is in its nucleus. Now we saw that the chemical properties of an atom are mainly determined by the number of electrons outside the largest complete shell. In addition they are *completely* determined by the *total* number of electrons. But there are as many protons in the nucleus as there are electrons outside it. Hence we can specify the chemical properties of an atom by stating the number of protons in its nucleus: this is known as its atomic number.

This atomic number must be distinguished from the atomic *weight* of an atom, which relates to the combined number of protons and neutrons in its nucleus (neglecting the weight of the electrons). Since neutrons and protons have practically the same weight, atomic weights will be *nearly* proportional to whole numbers, namely the total number of protons and neutrons in the nucleus.[2] For convenience, the unit of weight has been chosen to make the atomic weight of oxygen exactly equal to the number of particles in its nucleus. Since it contains 8 protons and 8 neutrons, its atomic number is 8 and its atomic weight 16. On this scale hydrogen, with a nucleus consisting of just 1 proton, has atomic number 1 and atomic weight 1.008.

This distinction between atomic weight and atomic number explains the existence of isotopes. Consider, for instance, the element carbon. It has 6 protons in its nucleus, so its atomic number is 6. But what of its atomic weight? this depends on how many neutrons there are in its nucleus. If there are 6

[2] We shall shortly meet another reason why the atomic weights are not exactly whole numbers.

neutrons its atomic weight is 12, if there are 7 neutrons it is 13. In either case the substance will be carbon, in the sense that its chemical properties will be just those conferred on it by the 6 electrons outside the nucleus. This, then, is an example of two isotopes—they have the same number of protons in their nucleus but different numbers of neutrons.

Since two isotopes have the same chemical properties, the elements we find in nature are usually mixtures of isotopes. Fortunately they can be separated into their constituents by processes which are sensitive to their difference in atomic weight. It is then found, for instance, that naturally occurring carbon is a mixture (of atomic weight 12.01) consisting of 99 per cent carbon 12 and 1 per cent carbon 13. This example is typical: most elements have one isotope whose abundance far outweighs the others. This is fortunate, since otherwise the first measurements of atomic weight would not have led to approximate whole numbers, and the elucidation of atomic structure would have been much more difficult.

The Abundances of the Elements

We are now in a position to appreciate the attempts that have been made to account for the existence in nature of the various elements. Why is the material of the universe distributed in this way, instead of all being, say, hydrogen, or some other simple substance? Moreover, the various elements are present in vastly different proportions—gold is much rarer than carbon, for instance. We can thus ask the more ambitious question: Why do the elements have their observed abundances?

The asking of this question was stimulated by the work of the American chemist F. W. Clarke (1847–1931), who appears to have been the first to seek a pattern in the observed abundances of the elements. From 1889 on he studied these abundances in the hope of finding a periodic relation between abundance and atomic weight, reminiscent of the periodic table. If such a relation exists, it would indicate that the abundance of an element depends on its chemical properties. The discovery of this relation would then help to explain the abundances. However, Clarke did not succeed in finding any pattern.

It will be noticed that he began his work before isotopes were discovered. Since, in fact, the different isotopes of an element have the same chemical properties but very different abundances, it is clear that Clarke's quest was hopeless—the abundances cannot be related to the chemical properties. Moreover, by using the improved data that subsequently became available, the American physicist W. D. Harkins discovered in 1917 that there *were* regularities in the abundances, but that these regularities were related to the properties of the *nuclei*. In particular, he found a correlation between the abundance of an element and the force needed to disrupt its nucleus. Harkins concluded that the key to the observed abundances must be sought not in chemistry but in nuclear physics.

The Structure of the Nucleus

The discovery that the problem of element formation is a problem in nuclear physics was the first big step forward. It makes our brief discussion of nuclear structure quite inadequate, so we shall now go into more detail. The key property of a nucleus is the one used by Harkins in his correlation, namely its ability to resist disruption. This prompts the fundamental question: What holds a nucleus together? It is certainly not gravitational forces, which are far too weak. Nor is it electrical forces, for neutrons are neutral, while the charged particles in the nucleus, the protons, actually repel one another.

This difficulty led physicists to introduce a new type of force into their description of nature. This so-called nuclear force acts between neutrons and neutrons, neutrons and protons, and protons and protons, binding these particles tightly together into the nucleus. It differs from gravitational and electrical forces in that it is a *short-range* force—that is, it is only appreciable when the two particles are very close together (Fig. 55). When this happens the nuclear force is strong enough

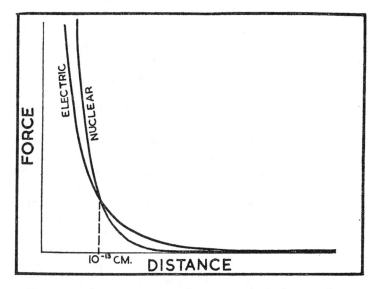

Fig. 55. The attractive nuclear force and the repulsive electric force between two protons. The nuclear force dominates at distances less than 10^{-13} centimeters. It is the same between two neutrons and between a neutron and a proton as between two protons.

to overcome the electrical repulsion between the protons, and so enables them to remain close together in the nucleus. The range of the nuclear force is actually about 10^{-13} centimeters, so that the size of a tightly bound nucleus will also be about 10^{-13} centimeters. By comparison, an atom as a whole is 100,000 times bigger (10^{-8} centimeters).

Since we are investigating the formation of particular nuclei in nature, it is important to understand what happens when a nucleus is formed as a result of the attractive forces between its constituents. We may compare this process with what occurs when a ball falls toward the earth from infinity. If the ball bounces elastically at the earth's surface it will rise at the same speed as it fell, and it will be able to move indefinitely away from the earth. On the other hand, it may happen that when the ball strikes the earth it loses energy as a result of the impact. In that case the ball will rise more slowly than it fell, and it may have less than the escape velocity. If so, it will remain bound to the earth. The energy lost by the ball will be converted into heat at the point of impact, and this heat will be removed by conduction, convection, and radiation.

In the same way, if protons and neutrons are brought close together by their attractive nuclear forces, they may also radiate some energy, and be left with insufficient to escape from each other's attraction. The result will be a set of protons and neutrons bound together, and this is just what a nucleus is. The amount of energy radiated away is known as the binding energy; it is also, of course, just the energy needed to disrupt the nucleus into its constituent particles. It was this binding energy which Harkins correlated with abundance. Its existence is also an additional reason why the atomic weights of nuclei are not exactly whole numbers. For, according to Einstein, the radiated energy carries off some mass, so that a set of protons and neutrons bound together in a nucleus have less mass than when they are far apart.

Now suppose a nuclear reaction takes place in which nuclei with a small binding energy are transformed into nuclei with a large binding energy. In such a reaction a large amount of energy must be radiated away. For instance, suppose two protons and two neutrons react together to form a nucleus—this will be a nucleus of helium, usually called an α particle. Now separated neutrons and protons have no binding energy, whereas an α particle is very tightly bound. As a result, a considerable amount of energy is released when an α particle is formed. How considerable can be appreciated from the fact that it is reactions like this which supply the energy radiated by the sun and stars, and which keep them hot. It is also reactions like this which it is hoped to exploit in fusion reactors as a cheap and virtually inexhaustible source of power.

Unfortunately, bringing protons and neutrons together to form α particles is not a trivial matter, because the protons repel one another electrically. This repulsion is stronger than the nuclear attraction, unless the protons are within the range of their nuclear forces. How can the protons be brought so close together that their nuclear forces will be dominant? The obvious answer is to have the protons moving so fast that their approach is not much hindered by electrical repulsion. In the laboratory these high velocities are produced by special accelerators—cyclotrons, synchrotrons, and so on. But how do the stars use nuclear reactions as their source of energy? They contain no cyclotrons. Clearly, the protons in stars must be moving fast without being specially accelerated. In other words, the stars must be *very hot,* if energy-producing reactions are to take place.

Nuclear reactions generated by high temperatures rather than by accelerators are known as thermonuclear reactions. One of the main tasks of fusion reactors is the creation of high temperatures, so that thermonuclear reactions can occur with a consequent production of energy. We shall see that it is in thermonuclear reactions that the key to our problem lies.

Thermonuclear Reactions

We must first understand in a general way which nuclei are likely to be formed by thermonuclear reaction. Suppose we begin with a system of protons and neutrons at a high temperature. Will larger and larger nuclei be built up until all the protons and neutrons are eventually in one vast nucleus? The answer is no. To understand what happens, we must say something more about the binding energy of nuclei.

We have seen that two nuclei can come together to form one nucleus as a result of the attractive forces between their constituent particles. When this occurs energy is radiated away, and the new nucleus is a bound unit that can be disrupted only if energy is added to it. When this was first stated, a proviso should have been added: the nuclei concerned must not be too large. This proviso is needed because in a large nucleus the protons on opposite sides are outside the range of their attractive nuclear forces, but are still repelling one another electrically. Such a nucleus is not very tightly bound. If it were to break up into two, the protons in each smaller nucleus would be closer together and more tightly bound by their attractive nuclear forces. In such a case, then, energy would be released by the *fission* of a nucleus. This is the process used in the first atomic bombs and in the existing nuclear power reactors.

It is important to realize that a large nucleus does not necessarily split up *spontaneously*. In order to become two nuclei, it may first have to pass through configurations which are less tightly bound than the original nucleus itself. In that case it can split up only if energy is first supplied, to enable it to adopt these intermediate configurations. This was the situation in the first atomic bombs; fission was induced by bombarding uranium and plutonium nuclei with neutrons. The resulting release of energy occurred with explosive force because a chain reaction was set up. When the uranium nucleus splits, it emits neutrons which can then induce fission in other uranium nuclei, and so on. In this way a small number of bombarding neutrons can trigger off a rapid release of energy. In a nuclear power reactor the rate of energy release is controlled by inserting impurities, which absorb many of the neutrons before they can induce fission.

The energy relations we have been describing can now be summarized in the following way. Small nuclei tend to fuse together and release energy. Large nuclei tend to split up and release energy. In order for either of these processes to occur, energy must first be borrowed, but the debt is soon repaid with interest.

We can now return to the question of which nuclei will be

formed as the result of a series of thermonuclear reactions. Both fusion and fission processes will occur, and a dynamic balance will be set up between them. A remarkable feature of this balance is that large nuclei are being continually formed and small nuclei are being continually split, although both these processes absorb energy. Reactions of this type are familiar to chemists under the name "endothermic." The necessary energy in this case is provided by the heat motion of the various nuclei. The amount of energy available for endothermic processes is therefore larger the hotter the system, so that the number of very small and very large nuclei is also larger the hotter the system.

Now suppose that, after a dynamic balance has been set up in a region of high temperature, this region is allowed to cool down slowly. We shall assume for simplicity that each time energy is released in a reaction it is radiated away and so lost to the system. This means that, when large nuclei split up and small nuclei fuse, the energy released is no longer available for inducing the reverse processes. So as the temperature slowly drops, the dynamic balance adjusts itself in such a way that the number of very small and very large nuclei steadily falls. Eventually the system will reach its most tightly bound condition, when no more energy can be released and its temperature is very low. When this occurs all the nuclei will be of intermediate size—too large to fuse further and too small to split further. This intermediate size corresponds to the iron nucleus. That is why we called iron rather than gold the perfect element to which nature proceeds. The universe aspires to the condition of iron.

The Abundances of the Elements

Our last conclusion immediately raises the question: Why does the universe not consist entirely of iron? Why are there both lighter and heavier elements in existence? One possibility is that the universe was formed with the elements in their observed abundances. However, we shall see that there are reasons for believing that the material of the universe was originally hydrogen. If this is so we could explain the present existence of elements *lighter* than iron simply by supposing that they have not yet had the opportunity to react together sufficiently.

The formation from hydrogen of elements heavier than iron is harder to explain. Their very existence shows that they must have participated in thermonuclear reactions—in regions hot enough for many endothermic processes to have occurred. What, then, prevented the resulting heavy elements from reverting to iron as the regions cooled? The answer is probably that the cooling was very *rapid,* or that an explosion occurred which rapidly threw the heavy nuclei out of the hot regions into their much cooler surroundings. In either case the dynamic balance has no time to adjust itself to the lower temperature. The heavy nuclei will then be stuck; there will be insufficient energy available to induce them to split. If this explanation is correct the existence of heavy nuclei is a manifestation of the dynamic balance that prevails at a high temperature; if the cooling is very rapid, this balance will be "frozen in" at low temperatures. This freezing helps to save us from an iron age.

Unfortunately, these complications make it impossible to calculate which nuclei will be formed, unless we know beforehand the precise history of the regions involved. We must therefore study the observed abundances in detail, in the hope that they contain useful clues to this history.

The main feature of the abundances is one that gradually emerged from various independent investigations, namely that they are, on the whole, the same everywhere in the Milky Way, so that we are really dealing with a cosmical phenomenon. The first hint of this came from studies of meteorites, which gave results very similar to those already established for the earth. Subsequent studies went further afield—the sun, the stars, the nebulae, and the interstellar gas were all explored.

The method of determining abundances in objects one cannot handle is to study their spectra. The abundances are obtained by measuring the darkness of the characteristic spectral lines produced by different elements. The detailed working out of this method is an extremely complicated and technical one. Nevertheless, many abundances have been established in distant objects, with the result that, for the most part, the abundances are the same everywhere in the Milky Way.[3]

There are, to be sure, certain detailed differences—some of which are of considerable importance. For instance, although hydrogen is by far the predominant element in general, there is relatively little of it on the earth. This particular fact can be easily explained in terms of the special circumstances obtaining at the earth, namely that its gravitational field is too weak to retain such a light gas, which is therefore able to escape. In general it appears to be possible to explain all local deviations from the universal abundances by special circumstances of this sort. Our main problem is thus cosmical rather than terrestrial.

The main features of the abundances which have to be explained are shown in Fig. 56. We can summarize them as follows. Ninety-three per cent of the atoms in the Milky Way are hydrogen, 7 per cent are helium, and only one atom in a thousand is a heavier element. This striking disparity between the abundances of hydrogen and the other elements is a valuable clue to the nature of what Ben Jonson called the remote imperfect matter which nature first begat. The obvious hypothesis is that matter started off as hydrogen—the simplest of all the elements—and that thermonuclear reactions occurred which led to the transformation of a small fraction of this material into heavier elements. In this view the earth itself is a mere impurity speck.

It is true that helium has an appreciable abundance relative to hydrogen, but we know in fact that helium is built up out of hydrogen in the centers of stars like the sun. As we mentioned, this is just the process that keeps them burning. Furthermore, we can calculate how much helium would be formed by all

[3] In this chapter we shall confine ourselves to the Milky Way.

the stars in the Milky Way, burning their fuel from the time they were formed until today. The calculation is only approximate because we do not know all the relevant facts very accurately, but the result is in rough agreement with the observed abundance of helium.

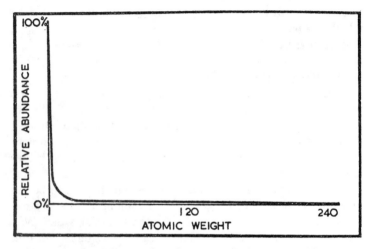

Fig. 56. The relative abundances of the elements. Ninety-three per cent of all atoms in the Milky Way are hydrogen, 7 per cent are helium, and one atom in a thousand is a heavier element. The actual curve fluctuates violently, and the diagram shows only its main trend.

Encouraged by this result, we naturally go on and try to build up the elements heavier than helium in the same way. Unfortunately, we immediately meet a snag. Heavier nuclei contain more protons, so that they exert a greater electrical repulsion on incoming protons. Higher speeds are then needed to overcome this repulsion, and this means higher temperatures. But the temperature in stars like the sun is only just high enough to build helium. Here we face a fundamental difficulty. This is how Eddington reacted to it in 1926:

It is held that the formation of heavy elements from hydrogen would not be appreciably accelerated at stellar temperatures, and must therefore be ruled out as a source of stellar energy. But the heavy elements which we handle must have been put together at some time and some place. We do not argue with the critic who urges that the stars are not hot enough for this purpose; we tell him to go and find a *hotter place*.

In the last few years Eddington's advice has been taken, and two main candidates for the hotter place have been proposed—one suggested by evolving models of the universe, and the other by the steady state model. We begin by describing the first possibility, popularly known as the α-β-γ theory.

α-β-γ Theory

According to this theory, the hot place desired by Eddington is nothing less than the whole universe a few minutes after the beginning of the expansion. The theory takes its name from its actual proposers, Ralph Alpher and George Gamow, and from its nominal proposer, the well-known nuclear physicist Hans Bethe. In 1948 they attempted to account for the abundances of the elements in terms of thermonuclear reactions occurring in the first half hour after the expansion began—at later times the universe would be too cool for these reactions to take place.

Unfortunately, this theory faces a serious difficulty. The helium nucleus (atomic weight 4) is very tightly bound, but the nucleus with one more particle in it (atomic weight 5) is not bound at all. It could exist for a very short while but would then emit one of its constituent particles. This absence of a stable nucleus of atomic weight 5 acts as a bottleneck to prevent the build-up of heavier elements. The bottleneck would be by-passed if two particles hit a helium nucleus simultaneously, but the density of particles is too small for this to happen often enough. Many attempts have been made to get around this bottleneck, but all without success. Indeed, Gamow himself wrote in 1956:

> Since the absence of any stable nucleus of atomic weight 5 makes it improbable that the heavier elements could have been produced in the first half hour in the abundances now observed, I would agree that the lion's share of the heavy elements may well have been formed later in the hot interiors of stars.

Gamow is here referring to the other theory of the hot place, namely that invoked by followers of the steady state model of the universe. This theory locates its hot place in the interiors of the relatively few stars which are substantially hotter than the sun.

Stellar Theory

Steady state theorists cannot, of course, invoke an early hot phase of the universe; for them the universe has always had the temperature it has now. On the other hand, the matter that is continually being created presumably appears in the form of hydrogen, the simplest of the elements. A hot place is therefore still needed to build up the heavy elements. The only hot places allowed are ones which can now be observed. It is therefore suggested that the heavy elements are built in unusually hot stars and are then distributed throughout the Milky Way by the explosions of varying degrees of violence which these stars are known to undergo. Of course the correctness of this suggestion would not in itself favor the steady state model, since the stellar processes envisaged could equally well occur in the evolving models. On the other hand, its incorrectness would be very embarrassing, and despite the bottleneck difficulty, would strongly support an evolving model of the universe.

There are several lines of evidence which suggest that at

least some of the heavy elements were formed since the birth of the Milky Way (which is believed to have taken place about 5 billion years ago). Perhaps the most intriguing comes from the discovery of the element technetium (atomic number 43) in certain stars. Now technetium is one of the "missing" elements—that is, it is an element which, according to the periodic table, should have existed but which had not in fact been found on the earth. In 1952 P. W. Merrill, an American astronomer, identified the spectrum of technetium in the so-called S stars. Its absence from the earth is not surprising, because we now know that all the isotopes of technetium are unstable—that is, their nuclei emit charged particles, the residual nuclei corresponding to other elements. The longest-lived isotope lasts less than a million years before emitting a particle, so if any technetium was present when the earth was formed 4 billion years ago, it has disappeared long since. Conversely, Merrill's discovery shows that some technetium must have been formed somewhere within the last million years.

This observational evidence encourages us to calculate how many elements are formed by the various known types of star, to see whether they can all be accounted for in this way. The calculations involve a fascinating combination of nuclear physics and the theories of stellar structure and stellar evolution. We shall describe only the modern version of these calculations, which is mainly due to Fred Hoyle.

The first step involves the formation of helium out of hydrogen in the centers of average stars like the sun. This process has been understood for some time. The more challenging part of the problem is to build the elements heavier than helium. To see how this can be done we ask the question: What happens to a star which has lived so long that it has burned most of the hydrogen in its central regions? This question naturally arises because stars more massive than the sun burn their fuel very quickly. The answer is that when its fuel is substantially depleted the star will cool down. As a result of this cooling the gas pressure inside the star decreases and is no longer able to hold the star out against its own gravitation. Consequently the star begins to contract.

This gravitational contraction releases a considerable amount of energy, so that the temperature rises again: a process akin to the earth being heated by a falling ball. We may note in passing that this process of heating by gravitational contraction was considered by the English physicist Lord Kelvin and the German Hermann von Helmholtz to be the source of the sun's energy. In fact, the energy produced in this way would not keep the sun burning for long enough,[4] but of course in the nineteenth century nothing was known of thermonuclear reactions.

As the star contracts, its temperature continues to rise, passes the original temperature at which the hydrogen was burned, and eventually becomes high enough for further thermonuclear reactions to occur. These new reactions release so much energy that the contraction is stopped. In the course of these reactions helium is converted into the various elements up to magnesium 24. There is no bottleneck at helium in this theory, because the density of the material inside a star is very much greater than in the universe as a whole in the α-β-γ theory. It is now quite a frequent occurrence for two particles to strike the helium nucleus simultaneously, so that it jumps the barrier at mass 5.

When all the helium fuel is burned up the cycle repeats itself. The star cools again and then begins to contract. As before, this contraction keeps heating up the star until more thermonuclear reactions can occur, the energy release of which stops the contraction. This time the elements are built up all the way to iron. The building-up process stops there because iron is the most stable of all the elements. As we explained earlier, elements heavier than iron will be built up only if an appreciable number of endothermic reactions take place. This needs higher temperatures.

[4] The supply of heat would last for only about 20 million years, whereas the sun is 300 times as old as that.

Hoyle believes that these high temperatures occur in the violently exploding stars known as supernovae, which at their peak outshine the galaxy that contains them. Almost immediately after the elements heavier than iron have been formed, they are ejected by the supernova explosion into the surrounding cold of space. This rapid cooling "freezes in" the heavy nuclei. They do not turn back into iron despite its greater stability, because to do so each nucleus would first have to pass through a configuration of *higher* energy. When the heavy elements are dispersed in space they have no means of obtaining this extra energy, so they just persist indefinitely.

Conclusions

The existence of the whole range of elements from helium to uranium can thus be traced to processes occurring in the various types of star. Many calculations will have to be performed before this stellar theory of element formation is established beyond doubt, but already some of its main features have been verified in detail. If the theory as a whole is correct we will have succeeded, as we did with the galaxies, in relating the contents of the universe to the laws of nature. It is true that the properties of the stars appear as accidental initial conditions, but this is not an essential part of the calculation. For if we adopt the steady state model, we have available a theory of galaxy formation which contains no arbitrary features, and so we should be able to deduce unambiguously that galaxies break up into the various known types of star. When this has been done we will have constructed a continuous chain of argument from the formation of the galaxies to the formation of the elements.

Satisfying as this will be, it does not necessarily mean that our task is finished—that the realm of the accidental can be diminished no further. We have still to see whether more detailed properties of the universe—such as the existence of planets or of life—are also inevitable consequences of the laws of nature. And if they *are* inevitable we must then accept the ultimate challenge: to show that we ourselves are not some haphazard outcome of the byplay of cosmic forces. For surely we are no more accidental than the matter we are made of.

EPILOGUE

THE UNITY AND THE UNIQUENESS OF THE UNIVERSE

> One is one and all alone, and ever more shall be so.
> POPULAR SONG

It will have become evident to the reader of this book that the exploration of the universe involves the study of a large number of problems, each with its own characteristic flavor. Perhaps the main difficulty in cosmology is to see all these problems in perspective, so that one has a clear grasp of their interrelations. I do not pretend that such understanding can come before the problems have been lived with, but perhaps a brief summary will help to clarify the situation. After this summary I shall indicate how the two main themes of this book—the unity and the uniqueness of the universe—may be related to one another.

Part I described how our present observational picture of the universe has been built up. After considerable controversy there is now almost universal agreement about the main outlines of this picture. The building blocks of the universe appear to be clusters of galaxies; a galaxy, such as our own Milky Way, being an organized collection of about 10 billion stars. A cluster, in turn, is an organized collection of galaxies, its contents ranging from two to several thousand members. These clusters are the building blocks of the universe in the sense that they do not appear to be grouped into superclusters but are more or less uniformly distributed throughout space.

The outstanding feature of this system of clusters is the famous "expansion of the universe," deduced from the Doppler shifts toward the red in the spectra of galaxies. These redshifts persist out to the greatest distances yet measured, about 7 billion light-years. They imply that clusters of galaxies are receding from one another, and at speeds proportional to their distances apart (Hubble's law). This proportionality is important, since it means that the expansion of the system would be described in the same way by every cluster. In other words, the expansion treats all clusters equally. This supports the idea that clusters are the building blocks of the universe.

One of the most important quantitative properties of the universe is the *rate* at which it expands. Suppose we were to assume that each cluster has always had the velocity we now observe it to have. Then at a certain time in the past all the clusters would have been crowded on top of one another. This time, which is called Hubble's constant, is about 10 billion years. Of course this simple extrapolation into the past may be quite incorrect; but it suggests that a time of 10 billion years is a significant one for cosmical phenomena. (The ages of the earth and sun are also of this order.) The most distant galaxies that can be observed by the 200-inch telescope emitted their light about 7 billion years ago, so that we have a glimpse of the universe as it was in the past. Unfortunately, this is not quite long enough ago for changes in its large-scale structure to be detectable, so that in practice we are seeing the universe at only one moment of its history. This frustrating situation may not last long, however, since radio telescopes can detect galaxies that are beyond the range of optical telescopes. The further development of radio astronomy may well bring with it the observational determination of the past history of the universe.

We now turn to theoretical considerations, about which, it must be admitted, there is as yet no general agreement. Our first major theme was the unity of the universe, which is a consequence of the appreciable forces acting between widely separated bodies. This unity implies that we can hope to understand any part of the universe (such as our own neighborhood) only by taking account of the whole universe. This dominating influence of the whole universe can be traced to its great bulk, which more than compensates for its great distance. The most important manifestation of this influence is the inertia of matter, which is a direct consequence of the gravitational forces exerted by distant galaxies. This example illustrates how the unity of the universe can be used to explain fundamental properties of matter.

We can also use the unity of the universe in the reverse direction. That is, we can infer from the behavior of our own neighborhood something about the properties of the universe at great distances. For instance, the observed amount of light in the night sky and the value of the gravitational constant enable us to estimate Hubble's constant and the average density of matter in the universe. Both these theoretical results refer to regions beyond the range of the optical telescope, so in this sense theory is a more powerful tool than observation for exploring the reaches of the universe. In fact the theoretical results for distant regions are in rough agreement with the observational results for nearby ones, a strong indication that the uniformity of the universe extends well beyond the range of existing observations.

Our second major theme, which was introduced independently of the first, was that the uniqueness of the universe raises special problems which are not encountered elsewhere in science. This was illustrated by contrasting the behavior of projectiles with that of the universe. The laws governing projectiles must be flexible enough to permit all directions and velocities to occur in practice. Moreover, no actual trajectory is of fundamental significance, only the properties common to all trajectories (such as being parabolic). It is the common properties of a multiplicity of phenomena that we are here concerned with—in this case, Newton's laws of motion.

By contrast, we do not have available many universes, some expanding, some contracting, some non-uniform, and so on, whose common properties can be established by observation and then enshrined in laws of nature. We have only one universe, so the significant fact in this case is the actual behavior of this single phenomenon. The laws of nature must thus be formulated in such a way that they relate only to the actual universe, for other universes, by definition, cannot exist. In other words, we seek a theory which describes all that actually happens, and nothing that does not, a theory in which everything that is not forbidden is compulsory.

Such a rigid theory has not yet been discovered. For instance, general relativity, which is the best theory of gravitation that has so far been proposed, is consistent with an infinite number of different possibilities, or models, for the history of the universe. Needless to say, not more than one of these models can be correct, so that the theory permits possibilities that are not realized in nature. Fortunately we may possess a clue to a rigid theory, since these models are themselves not equally rigid.

The main contrast is between the steady state model on the one hand and the evolving models on the other. According to the steady state model, the large-scale structure of the universe does not change with time: although the universe is expanding, its density can remain constant because new matter is being continually created. In the evolving models there is no creation; the universe becomes continually more dilute, having exploded from a highly condensed state about 10 billion years ago. The observations described in Part I are not sufficiently precise to rule out either type of model, but only the steady state one leads to a unique set of possible values for various properties of the universe (such as the average mass of a galaxy, or the abundances of the elements). This suggests that the rigid theory of gravitation we are seeking should imply that the universe *is* in a steady state.

We do not yet know how to construct such a theory, or any other rigid theory for that matter, but I strongly suspect that to do so we shall need to exploit the unity of the universe. For only if the different parts of the universe are related in theory can we expect to account for the actual relations that are observed in practice. In this way our two main themes become linked together.

A possible linkage of this sort has been suggested by Bondi and Gold. They begin by making the obvious point that the large-scale behavior of the universe depends on the laws of motion. They then add that, owing to the unity of the universe, the laws of motion depend on its large-scale behavior. Such a self-linked system may lead to only one possibility for the universe, since it faces serious problems of self-consistency. Suppose, for example, we were to adopt some particular model for the universe. Owing to the unity of the universe, this would have certain implications for the laws of motion, which in turn would restrict the large-scale behavior of the universe. Now these restrictions may be inconsistent with the model that was originally adopted, in which case this model is ruled out. Bondi and Gold suggested that the requirement of self-consistency might be so severe as to permit only one model of the universe. If they are right, the unity and the uniqueness of the universe are linked together in the most intimate way, the universe itself being the sole self-perpetuating consequence of its long-range interactions. The discovery of all these interactions, and the establishment of their unique significance, would be a fitting climax to man's long struggle to understand the universe.

INDEX

α-β-γ theory, 195
Absolute space: and motion, 93–95; Berkeley's objection to, 97–98
Abundances of elements, 191 ff., Fig. 56
Acceleration, absolute, 93; and inertia, 84 ff.; and Mach's Principle, 83–85
"Accident" in universe, 164, 167, 173, 179–80
Adams, W. S., 23–24
Alpha Centauri, 21
Alpher, R., 195
Andromeda galaxy, 59, 60, 151. See Pl. IX
Apparent brightness of stars, Olbers' Paradox and, 71–78. See also Intrinsic brightness, Figs. 28–30
Aristarchus of Samos, 8, 9, 10, 11, 13; and moon's distance (with Hipparchus), 8–10
Aristotle, 5
Artificial satellites, 86, 91
Atoms: number and weight, 184–85; outer electrons of, 183; Rutherford-Bohr model, 182, Fig. 54; structure of, 181

Baade, W., 51, 52
Baum, W., 63, 64
Berkeley, Bishop, 97–98
Bessell, F. W., 17, 18, 19, 21
Bethe, H., 195
Bode's Law, 164, 165, 182
Bondi, H., 158, 205
Brightness of stars, 18–24; apparent, 72–76

Caille, Abbé, N. L. de la, 12
Cassirer, E., 100
Cavendish, H., 123, 124
Cepheids I and II, 51–52; and distance determination, 24–25, 35, 36, 49, 50, 51–52; period-luminosity, relations for, 25, 27, Figs. 12, 22; variables, 24, 25, 49–50
Charlier, C. V. L., 79, Fig. 32
Christiansen, W. N., 44
Clarke, Agnes, 48
Clarke, F. W., 185–86
Clock and gravitation, 136–38
Clock-paradox, 131–38

207

INDEX

Clusters (stars and galaxies), formation of, 177–80; open, 40–42, Fig. 18; unresolved, 48. See also Pls. VI, VIII
Colliding galaxies, 65, Pl. XIII
Copernicus, 52
Coriolis forces, 92, 95, 104, 125–27, Figs. 38, 45
Cosmological Letters (Lambert), 33
Cosmological principle, 153–54; perfect, 154, 159
Cosmologist and physicist, compared, 166–67
Coulomb's Law, 116, 117, 118, 119
Creation of the Universe, The (Gamow), 172
Curtis, H. D., 49
Curvature of space, 141–46

Darkness at night, 80
Delta Cephei (star), 24, Fig. 11
Donne, John, 104
Doppler, C., 56
Doppler effect, 38, 56–58, 121–22, 133, 134, Figs. 23, 46
Doppler shifts of galaxies, 63, 80, 202

Eddington, Sir A., 48, 49, 59, 61, 113, 114, 124, 194, 195
Einstein's theories, 85, 99, 107, 109, 111, 112, 113, 115, 117, 131, 132, 188; general theory of relativity, 115–16, 139–50, 204; applied to universe, 155 ff.; field equations of, 146–50

Electric forces, 116–18, 187–88
Electromagnetism, 116, Fig. 20
Electrons, outer, of atom, 183
Elements: abundances of, 185–86. See also *Period Table*, formation of, 181 ff.
Endothermic reactions, 191
Equivalence principle, 107–14; Einstein's explanation of, 111; formulation of, 111; Sir E. Whittaker on, 113–14
Eratosthenes, 5–6, 8, Fig. 1
Euler, L., 98, 101
Evolving models of universe, 171–73
Ewen, H. I., 44
Expansion of universe, 55–65, 201–5; discovery of, 55–65; rate of, 80, 202

Field equations, Einstein's, 146–50
Fizeau, H., 58
"Flying Star, The," 19
Fog, interstellar, 40, 41, 42, 48, Fig. 19
Forces, centrifugal, 89, 95, 98–99, 104–5, 125–27, Fig. 36; Coriolis, 92, 95, 104, 125–27, Figs. 38, 45; electrical and magnetic, 109; gravitational, 85, 93, 105, 107, 108, 111–13, 115 ff.; gravitational and inertial, 107, 108, 109 ff., 121–22. See also Figs. 48, 50
Foucault, L., 95
Foucault pendulum, 95, 99, 102, 104, Fig. 40, Pl. XIV
Friedmann, A., 157

INDEX

Fundamental and accidental features, 70, 163–64, 167, 173, 199–200

Galaxies: basic properties of, 170–71; and clusters, uniform distribution of, 64; colliding, 65, Pl. XIII; conclusions concerning, 199–200; double, 177; external, 34, 47–53; formation of, 167–80; Hubble on, 49–53, Pl. XI; in space (distribution), 51; local group, 50; Milky Way as spiral galaxy, 43–44; three types of, 51; self-propagation of, 174. See also Fig. 52 for birth of
Galilei, Galileo, 3, 84, 108, 111, 144, 162
Gamow, G., 172, 195, 196
Gauss, C. F., 116
General relativity, 139–50, 204
Globular clusters, 40–42, Pl. VI
Gold, T., 158, 205
Goodricke, J., 24
Gravitation: clock and, 136–37; constant (gravitational), 122–25; force (gravitational) of accelerating stars, 119–22; simple theory of, 116–19. See also Figs. 41, 42, 48
Greater Magellanic cloud, Pl. IV
Greek astronomers, 4, 15 ff.; star-distance estimates by, 15

Halley, E., 13
Harkins, W. D., 186, 188
Helium, 188, 194, 198, 199; formation of, from hydrogen, 197, 198; nucleus and α-β-γ theory, 195
Helmholtz, H. L. von, 198
Henderson, W., 19
Herschel, Sir J., 21, 34, 35, 41
Herschel, Sir W., 34, 36, Fig. 14
Hindman, 44
Hipparchus, 19. See also Aristarchus, Greek astronomers
Hoyle, F., 158, 197, 199
Hubble, E. P., 49, 52, 55, 62, 63, 151; Hubble's constant, 63, 122, 124, 178 n.; Hubble's Law, 64, 64 n., 202. See also Pl. XI
Huggins, Sir W., 58
Hulst, H. C. van de, 43
Humason, M. L., 63
Huygens, C., 15, 16, 17
Hydrogen: and helium, 197–98; interstellar clouds of, 43–44; as original material of universe, 192; spectrum of, 23–24, 43; and supergiants, 44

Inertia: Berkeley on, 97–98; Eddington and Whittaker on, 113–14; Einstein on (see Einstein's theories); Galileo on, 84; inertial and non-inertial frames, 88, 99; and influence of distant stars, Chs. VIII and IX *passim*; Mach's approach to, 98–105; Newton on intrinsic nature of, 84–97; origin of, 115–29. See also Figs. 39 ff.
Inertial conditions, 162, 166
Interstellar fog, 40–42; and Olbers' Paradox, 76

INDEX

Interstellar hydrogen clouds, 44
Intrinsic and apparent brightness, 23, 40–42, 74–76
Inverse square law: Coulomb's, 116, 117–18, 119; Newton's, 116, 117–18, 146; in Olbers' Paradox, 74, Fig. 7
Iron, as perfect element, 181, 191–92
Island-universe theory, 48
Isotopes, 184, 185, 186

Kant, I., 29, 31–32, 34, 37, 43, 47
Kapteyn, J. C., 35, 36, 37, 40, 41
Kelvin, Lord, 198
Kohlschutter, A., 23

Lagrange, J. L., 3
Lambert, J., 30, 34
Leavitt, Henrietta, 25
Lemaître, Abbé G., 157
Lesser Magellanic cloud, 25, Pl. III
Leverrier, U. J., 116
Light: and distance (see Doppler effect); as electromagnetic waves, 116; frequency of and distance (see Doppler effect); hours, years, 14; photons, 72, 73, 74, 80
Lindblad, B., 37
Lowell Observatory, 59
"Lunik," USSR launches, 91 n.
Lyttleton, R. A., 165

Mach, E., 85, 123, 125, 150 n.; Mach's Principle, 83–85, 112, 113, 115; criticism of, 100–2
Mendeleef, D. I., 182

Mercury, the planet, 117, 147, 164
Merrill, P. W., 197
Milky Way, 29–34, 61, 62, 128, and *passim;* abundances of elements in, 191 ff.; age of, 178; Baade's discovery of, 52; fog belt in, 40, 41, 48–49; rotation of, 37; as spiral galaxy, 43–44; stars rotating around center of (1926 discovery), 104; models of, 29 ff. See also Figs. 13 ff. and Pls. V, VII
Milne, E. A., 155
Models of universe, evolving and steady state, 171–76
Moon: ancient and modern ideas on distance of, 7–13; variations in, 8 n.; partial eclipse of, 8
Morgan, W. W., 43, 44
Motion: absolute, 93–94; Newton's second law of, 85–88, 101; proper, 19; relative, Berkeley on, 97–98; Mach on, 98–99
Mount Palomar, 63
Mount Wilson, 36
Müller, K., 44

Nebulae: as external galaxies, 47–49; spiral, 59, Pls. IX, X
Neptune, the planet, 165
Newton, Sir I., 3, 17, 18, 23, 84, 85, 101, 103, 104, 107, 116, 117, 119, 122, 123, 125, 128, 131, 135, 146, 147, 162, 163, 165; on inertia, 84–97;

INDEX

Newton, Sir I. (cont'd)
  inverse square law of, 116, 117, 118, 146; second law of motion of, 85–88, 101, 203; *Principia* of, 93–95. See also Fig. 34
Non-Euclidean geometry, 141–46
Novae, 49
Nuclear fission, 189 ff.
Nuclear forces, 186–89
Nuclear fusion, 188–91
Nucleus, structure of, 186–89

Olbers, H., 71 ff.; Olbers' Paradox, 71–80, 120, 121, 127, 129, 156; resolution of Paradox, 76–80
Oort, J. H., 37, 38, 42, 44
Open star clusters, 31–33, Pl. VIII
Orion, 33, 48

Parabola explained, 162–63, Fig. 51
Parallax concept in astronomy, 12; and star distance measurements, 15 ff. See esp. Figs. 5–8
Perfect cosmological principle, 153–54, 158
Periodic table of elements, 182, 185
Period-luminosity relation (cepheids), 25–27, 49, 50, 51, Figs. 12, 22
Photons, 72, 80, 136 n., Fig. 33
Planets, measurements of distance of, from sun, 14; Bode's Law (Bode's relation) and, 164–65
Pluto, the planet, 14, 165
Pole Star, 25
Principia (Newton), 93–97
Principles of Mathematics, The (Russell), 100
Projectiles, motion of, 144, 162, 163, 164, 203
Proper motion, 19 ff., Fig. 10
Protons, neutrons, electrons compared, 184, Fig. 55
Ptolemy, 19
Purcell, E. M., 44

Radial velocities: bright stars, 55 ff.; galaxies, 55, 59 ff.
Radio astronomy, 64 ff.
Realm of the Nebulae (Hubble), 50, 63
Reds shifts. See Doppler shifts, Pl. XII
Relativity, Einstein's general theory of, 115, 139–50, 204
Robertson, H. P., 155
Rosse, Lord, 43, 47
Rotation: absolute (Newton), 93–95; differential and rigid compared, 38; Foucault's demonstration of, 95; Mach's demonstration of, 104. See also Mach, Figs. 16, 17
Russell, Bertrand, 100, 102, 103, 161
Rutherford-Bohr model of atom, 182, Fig. 54

S stars, 197
Sagittarius, 36, 38. See also Pl. VII

Shapley, H., 36, 38, 40, 42, 52, Fig. 15
Sirius, 58, 102; Huygens' estimate of distance of, 15; Newton's, 17
Sitter, W. de, 157
Slipher, V. M., 59, 61, 62
Solar system, motions of, 37–38, 163–66; size of, 5–14
Space, curvature of, 141–46
Spectrum: electromagnetic, 44, Fig. 20; of hydrogen, 44; laboratory, 45; spectral lines and Doppler effect, 58; of stars, 24–25; of sun, Pl. II
Spiral galaxies, 43 ff.; radial velocities of, 55, 59 ff.; and 21-cm astronomy, 43–44
Stars: apparent brightness of, 72–76 (*see also* Intrinsic brightness); average distance between, 77; distance to, 15–27; indirect method of measuring, 23–27; spectrum of, 23–25; supergiant, 43–44, 50. *See also* Clusters, Galaxies, etc., *and* named stars
Steady state theory, 173–76, 196; conclusions concerning, 179–80
Stellar parallax, 17–23
Stellar theory of element formation, 196–99
Struve, O., 21
Sun: calculation of diameter of, 10, 11, Fig. 4; corona of, Pl. I; corona of, at eclipse, 10; gravitational force of, 85–93, Fig. 34; Greek measurements of distance of, 10–11; modern measurements of, 13; relation of Milky Way to, 38–40; spectrum of, Pl. II
Supergiant stars, 43–44, 50; and hydrogen, 44
Supernovae, explosions of, 199

Technetium (element) in certain stars, 197
Thermonuclear reactions, 189–95
Thomson, Sir J. J., 183
Tisserand, F., 116–17
Titius, J. D., 164
Trumpler, R. J., 40, 41, 48
21-cm astronomy, 43–44, 65

Uniqueness of universe, vii ff., 161–67, 201–5
Universe: contents of, 166–67; expansion of, 55–64, 201–5, Fig. 26; history of, 151–60; initial conditions of, 161–66; uniqueness of, vii ff., 161–67, 201–5
Unresolved and resolved star clusters, 48
Uranium, 182, 199
Uranus, 165

Vega, 21
Velocities: and distances of galaxies (Hubble), 62–64, Fig. 25; total, radial, and transverse, 19, 38, 55 ff. *See also* Figs. 9, 24, 25
Virgo constellation, 50
Vistas in Astronomy (ed. A. Beer), 159 n.

Walker, A. G., 155
Weber, W., 116
Whitehead, A. N., 102
Whittaker, Sir E., 113, 114, 124

Wright, T., 29–31, 34, Fig. 13

Yaplee, B. S., 13

CATALOG OF DOVER BOOKS

Astronomy

BURNHAM'S CELESTIAL HANDBOOK, Robert Burnham, Jr. Thorough guide to the stars beyond our solar system. Exhaustive treatment. Alphabetical by constellation: Andromeda to Cetus in Vol. 1; Chamaeleon to Orion in Vol. 2; and Pavo to Vulpecula in Vol. 3. Hundreds of illustrations. Index in Vol. 3. 2,000pp. 6$1/8$ x 9$1/4$.
Vol. I: 0-486-23567-X
Vol. II: 0-486-23568-8
Vol. III: 0-486-23673-0

EXPLORING THE MOON THROUGH BINOCULARS AND SMALL TELESCOPES, Ernest H. Cherrington, Jr. Informative, profusely illustrated guide to locating and identifying craters, rills, seas, mountains, other lunar features. Newly revised and updated with special section of new photos. Over 100 photos and diagrams. 240pp. 8$1/4$ x 11. 0-486-24491-1

THE EXTRATERRESTRIAL LIFE DEBATE, 1750–1900, Michael J. Crowe. First detailed, scholarly study in English of the many ideas that developed from 1750 to 1900 regarding the existence of intelligent extraterrestrial life. Examines ideas of Kant, Herschel, Voltaire, Percival Lowell, many other scientists and thinkers. 16 illustrations. 704pp. 5$3/8$ x 8$1/2$. 0-486-40675-X

THEORIES OF THE WORLD FROM ANTIQUITY TO THE COPERNICAN REVOLUTION, Michael J. Crowe. Newly revised edition of an accessible, enlightening book re-creates the change from an earth-centered to a sun-centered conception of the solar system. 242pp. 5$3/8$ x 8$1/2$. 0-486-41444-2

ARISTARCHUS OF SAMOS: The Ancient Copernicus, Sir Thomas Heath. Heath's history of astronomy ranges from Homer and Hesiod to Aristarchus and includes quotes from numerous thinkers, compilers, and scholasticists from Thales and Anaximander through Pythagoras, Plato, Aristotle, and Heraclides. 34 figures. 448pp. 5$3/8$ x 8$1/2$.
0-486-43886-4

A COMPLETE MANUAL OF AMATEUR ASTRONOMY: TOOLS AND TECHNIQUES FOR ASTRONOMICAL OBSERVATIONS, P. Clay Sherrod with Thomas L. Koed. Concise, highly readable book discusses: selecting, setting up and maintaining a telescope; amateur studies of the sun; lunar topography and occultations; observations of Mars, Jupiter, Saturn, the minor planets and the stars; an introduction to photoelectric photometry; more. 1981 ed. 124 figures. 25 halftones. 37 tables. 335pp. 6$1/2$ x 9$1/4$. 0-486-42820-8

AMATEUR ASTRONOMER'S HANDBOOK, J. B. Sidgwick. Timeless, comprehensive coverage of telescopes, mirrors, lenses, mountings, telescope drives, micrometers, spectroscopes, more. 189 illustrations. 576pp. 5$5/8$ x 8$1/4$. (Available in U.S. only.)
0-486-24034-7

STAR LORE: Myths, Legends, and Facts, William Tyler Olcott. Captivating retellings of the origins and histories of ancient star groups include Pegasus, Ursa Major, Pleiades, signs of the zodiac, and other constellations. "Classic."—Sky & Telescope. 58 illustrations. 544pp. 5$3/8$ x 8$1/2$. 0-486-43581-4

CATALOG OF DOVER BOOKS

Chemistry

THE SCEPTICAL CHYMIST: THE CLASSIC 1661 TEXT, Robert Boyle. Boyle defines the term "element," asserting that all natural phenomena can be explained by the motion and organization of primary particles. 1911 ed. viii+232pp. 5³/₈ x 8¹/₂.
0-486-42825-7

RADIOACTIVE SUBSTANCES, Marie Curie. Here is the celebrated scientist's doctoral thesis, the prelude to her receipt of the 1903 Nobel Prize. Curie discusses establishing atomic character of radioactivity found in compounds of uranium and thorium; extraction from pitchblende of polonium and radium; isolation of pure radium chloride; determination of atomic weight of radium; plus electric, photographic, luminous, heat, color effects of radioactivity. ii+94pp. 5³/₈ x 8¹/₂. 0-486-42550-9

CHEMICAL MAGIC, Leonard A. Ford. Second Edition, Revised by E. Winston Grundmeier. Over 100 unusual stunts demonstrating cold fire, dust explosions, much more. Text explains scientific principles and stresses safety precautions. 128pp. 5³/₈ x 8¹/₂. 0-486-67628-5

MOLECULAR THEORY OF CAPILLARITY, J. S. Rowlinson and B. Widom. History of surface phenomena offers critical and detailed examination and assessment of modern theories, focusing on statistical mechanics and application of results in mean-field approximation to model systems. 1989 edition. 352pp. 5³/₈ x 8¹/₂. 0-486-42544-4

CHEMICAL AND CATALYTIC REACTION ENGINEERING, James J. Carberry. Designed to offer background for managing chemical reactions, this text examines behavior of chemical reactions and reactors; fluid-fluid and fluid-solid reaction systems; heterogeneous catalysis and catalytic kinetics; more. 1976 edition. 672pp. 6¹/₈ x 9¹/₄. 0-486-41736-0 $31.95

ELEMENTS OF CHEMISTRY, Antoine Lavoisier. Monumental classic by founder of modern chemistry in remarkable reprint of rare 1790 Kerr translation. A must for every student of chemistry or the history of science. 539pp. 5³/₈ x 8¹/₂. 0-486-64624-6

MOLECULES AND RADIATION: An Introduction to Modern Molecular Spectroscopy. Second Edition, Jeffrey I. Steinfeld. This unified treatment introduces upper-level undergraduates and graduate students to the concepts and the methods of molecular spectroscopy and applications to quantum electronics, lasers, and related optical phenomena. 1985 edition. 512pp. 5³/₈ x 8¹/₂. 0-486-44152-0

A SHORT HISTORY OF CHEMISTRY, J. R. Partington. Classic exposition explores origins of chemistry, alchemy, early medical chemistry, nature of atmosphere, theory of valency, laws and structure of atomic theory, much more. 428pp. 5³/₈ x 8¹/₂. (Available in U.S. only.) 0-486-65977-1

GENERAL CHEMISTRY, Linus Pauling. Revised 3rd edition of classic first-year text by Nobel laureate. Atomic and molecular structure, quantum mechanics, statistical mechanics, thermodynamics correlated with descriptive chemistry. Problems. 992pp. 5³/₈ x 8¹/₂.
0-486-65622-5

ELECTRON CORRELATION IN MOLECULES, S. Wilson. This text addresses one of theoretical chemistry's central problems. Topics include molecular electronic structure, independent electron models, electron correlation, the linked diagram theorem, and related topics. 1984 edition. 304pp. 5³/₈ x 8¹/₂. 0-486-45879-2

CATALOG OF DOVER BOOKS

Engineering

DE RE METALLICA, Georgius Agricola. The famous Hoover translation of greatest treatise on technological chemistry, engineering, geology, mining of early modern times (1556). All 289 original woodcuts. 638pp. 6¾ x 11. 0-486-60006-8

FUNDAMENTALS OF ASTRODYNAMICS, Roger Bate et al. Modern approach developed by U.S. Air Force Academy. Designed as a first course. Problems, exercises. Numerous illustrations. 455pp. 5⅜ x 8½. 0-486-60061-0

DYNAMICS OF FLUIDS IN POROUS MEDIA, Jacob Bear. For advanced students of ground water hydrology, soil mechanics and physics, drainage and irrigation engineering and more. 335 illustrations. Exercises, with answers. 784pp. 6⅛ x 9¼. 0-486-65675-6

THEORY OF VISCOELASTICITY (SECOND EDITION), Richard M. Christensen. Complete consistent description of the linear theory of the viscoelastic behavior of materials. Problem-solving techniques discussed. 1982 edition. 29 figures. xiv+364pp. 6⅛ x 9¼. 0-486-42880-X

MECHANICS, J. P. Den Hartog. A classic introductory text or refresher. Hundreds of applications and design problems illuminate fundamentals of trusses, loaded beams and cables, etc. 334 answered problems. 462pp. 5⅜ x 8½. 0-486-60754-2

MECHANICAL VIBRATIONS, J. P. Den Hartog. Classic textbook offers lucid explanations and illustrative models, applying theories of vibrations to a variety of practical industrial engineering problems. Numerous figures. 233 problems, solutions. Appendix. Index. Preface. 436pp. 5⅜ x 8½. 0-486-64785-4

STRENGTH OF MATERIALS, J. P. Den Hartog. Full, clear treatment of basic material (tension, torsion, bending, etc.) plus advanced material on engineering methods, applications. 350 answered problems. 323pp. 5⅜ x 8½. 0-486-60755-0

A HISTORY OF MECHANICS, René Dugas. Monumental study of mechanical principles from antiquity to quantum mechanics. Contributions of ancient Greeks, Galileo, Leonardo, Kepler, Lagrange, many others. 671pp. 5⅜ x 8½. 0-486-65632-2

STABILITY THEORY AND ITS APPLICATIONS TO STRUCTURAL MECHANICS, Clive L. Dym. Self-contained text focuses on Koiter postbuckling analyses, with mathematical notions of stability of motion. Basing minimum energy principles for static stability upon dynamic concepts of stability of motion, it develops asymptotic buckling and postbuckling analyses from potential energy considerations, with applications to columns, plates, and arches. 1974 ed. 208pp. 5⅜ x 8½. 0-486-42541-X

BASIC ELECTRICITY, U.S. Bureau of Naval Personnel. Originally a training course; best nontechnical coverage. Topics include batteries, circuits, conductors, AC and DC, inductance and capacitance, generators, motors, transformers, amplifiers, etc. Many questions with answers. 349 illustrations. 1969 edition. 448pp. 6½ x 9¼. 0-486-20973-3

CATALOG OF DOVER BOOKS

Mathematics

FUNCTIONAL ANALYSIS (Second Corrected Edition), George Bachman and Lawrence Narici. Excellent treatment of subject geared toward students with background in linear algebra, advanced calculus, physics and engineering. Text covers introduction to inner-product spaces, normed, metric spaces, and topological spaces; complete orthonormal sets, the Hahn-Banach Theorem and its consequences, and many other related subjects. 1966 ed. 544pp. $6^1/_8$ x $9^1/_4$. 0-486-40251-7

DIFFERENTIAL MANIFOLDS, Antoni A. Kosinski. Introductory text for advanced undergraduates and graduate students presents systematic study of the topological structure of smooth manifolds, starting with elements of theory and concluding with method of surgery. 1993 edition. 288pp. $5^3/_8$ x $8^1/_2$. 0-486-46244-7

VECTOR AND TENSOR ANALYSIS WITH APPLICATIONS, A. I. Borisenko and I. E. Tarapov. Concise introduction. Worked-out problems, solutions, exercises. 257pp. $5^5/_8$ x $8^1/_4$. 0-486-63833-2

AN INTRODUCTION TO ORDINARY DIFFERENTIAL EQUATIONS, Earl A. Coddington. A thorough and systematic first course in elementary differential equations for undergraduates in mathematics and science, with many exercises and problems (with answers). Index. 304pp. $5^3/_8$ x $8^1/_2$. 0-486-65942-9

FOURIER SERIES AND ORTHOGONAL FUNCTIONS, Harry F. Davis. An incisive text combining theory and practical example to introduce Fourier series, orthogonal functions and applications of the Fourier method to boundary-value problems. 570 exercises. Answers and notes. 416pp. $5^3/_8$ x $8^1/_2$. 0-486-65973-9

COMPUTABILITY AND UNSOLVABILITY, Martin Davis. Classic graduate-level introduction to theory of computability, usually referred to as theory of recurrent functions. New preface and appendix. 288pp. $5^3/_8$ x $8^1/_2$. 0-486-61471-9

AN INTRODUCTION TO MATHEMATICAL ANALYSIS, Robert A. Rankin. Dealing chiefly with functions of a single real variable, this text by a distinguished educator introduces limits, continuity, differentiability, integration, convergence of infinite series, double series, and infinite products. 1963 edition. 624pp. $5^3/_8$ x $8^1/_2$. 0-486-46251-X

METHODS OF NUMERICAL INTEGRATION (SECOND EDITION), Philip J. Davis and Philip Rabinowitz. Requiring only a background in calculus, this text covers approximate integration over finite and infinite intervals, error analysis, approximate integration in two or more dimensions, and automatic integration. 1984 edition. 624pp. $5^3/_8$ x $8^1/_2$. 0-486-45339-1

INTRODUCTION TO LINEAR ALGEBRA AND DIFFERENTIAL EQUATIONS, John W. Dettman. Excellent text covers complex numbers, determinants, orthonormal bases, Laplace transforms, much more. Exercises with solutions. Undergraduate level. 416pp. $5^3/_8$ x $8^1/_2$. 0-486-65191-6

RIEMANN'S ZETA FUNCTION, H. M. Edwards. Superb, high-level study of landmark 1859 publication entitled "On the Number of Primes Less Than a Given Magnitude" traces developments in mathematical theory that it inspired. xiv+315pp. $5^3/_8$ x $8^1/_2$. 0-486-41740-9

CATALOG OF DOVER BOOKS

Physics

OPTICAL RESONANCE AND TWO-LEVEL ATOMS, L. Allen and J. H. Eberly. Clear, comprehensive introduction to basic principles behind all quantum optical resonance phenomena. 53 illustrations. Preface. Index. 256pp. 5⅜ x 8½. 0-486-65533-4

QUANTUM THEORY, David Bohm. This advanced undergraduate-level text presents the quantum theory in terms of qualitative and imaginative concepts, followed by specific applications worked out in mathematical detail. Preface. Index. 655pp. 5⅜ x 8½.
0-486-65969-0

ATOMIC PHYSICS (8th EDITION), Max Born. Nobel laureate's lucid treatment of kinetic theory of gases, elementary particles, nuclear atom, wave-corpuscles, atomic structure and spectral lines, much more. Over 40 appendices, bibliography. 495pp. 5⅜ x 8½.
0-486-65984-4

A SOPHISTICATE'S PRIMER OF RELATIVITY, P. W. Bridgman. Geared toward readers already acquainted with special relativity, this book transcends the view of theory as a working tool to answer natural questions: What is a frame of reference? What is a "law of nature"? What is the role of the "observer"? Extensive treatment, written in terms accessible to those without a scientific background. 1983 ed. xlviii+172pp. 5⅜ x 8½.
0-486-42549-5

AN INTRODUCTION TO HAMILTONIAN OPTICS, H. A. Buchdahl. Detailed account of the Hamiltonian treatment of aberration theory in geometrical optics. Many classes of optical systems defined in terms of the symmetries they possess. Problems with detailed solutions. 1970 edition. xv + 360pp. 5⅜ x 8½. 0-486-67597-1

PRIMER OF QUANTUM MECHANICS, Marvin Chester. Introductory text examines the classical quantum bead on a track: its state and representations; operator eigenvalues; harmonic oscillator and bound bead in a symmetric force field; and bead in a spherical shell. Other topics include spin, matrices, and the structure of quantum mechanics; the simplest atom; indistinguishable particles; and stationary-state perturbation theory. 1992 ed. xiv+314pp. 6⅛ x 9¼. 0-486-42878-8

LECTURES ON QUANTUM MECHANICS, Paul A. M. Dirac. Four concise, brilliant lectures on mathematical methods in quantum mechanics from Nobel Prize-winning quantum pioneer build on idea of visualizing quantum theory through the use of classical mechanics. 96pp. 5⅜ x 8½. 0-486-41713-1

THIRTY YEARS THAT SHOOK PHYSICS: THE STORY OF QUANTUM THEORY, George Gamow. Lucid, accessible introduction to influential theory of energy and matter. Careful explanations of Dirac's anti-particles, Bohr's model of the atom, much more. 12 plates. Numerous drawings. 240pp. 5⅜ x 8½. 0-486-24895-X

ELECTRONIC STRUCTURE AND THE PROPERTIES OF SOLIDS: THE PHYSICS OF THE CHEMICAL BOND, Walter A. Harrison. Innovative text offers basic understanding of the electronic structure of covalent and ionic solids, simple metals, transition metals and their compounds. Problems. 1980 edition. 582pp. 6⅛ x 9¼.
0-486-66021-4

CATALOG OF DOVER BOOKS

ON ELECTRICITY AND MAGNETISM, James Clerk Maxwell. ...dation work of modern physics. Brings to final form Maxwell's theory of ...ism and rigorously derives his general equations of field theory. 1,084pp. ...vo-vol. set. Vol. I: 0-486-60636-8 Vol. II: 0-486-60637-6

...ATICS FOR PHYSICISTS, Philippe Dennery and Andre Krzywicki. Superb ...les math needed to understand today's more advanced topics in physics and ...ng. Theory of functions of a complex variable, linear vector spaces, much more. ...s. 1967 edition. 400pp. 6½ x 9¼. 0-486-69193-4

...DUCTION TO QUANTUM MECHANICS WITH APPLICATIONS TO ...MISTRY, Linus Pauling & E. Bright Wilson, Jr. Classic undergraduate text by Nobel ...winner applies quantum mechanics to chemical and physical problems. Numerous ...s and figures enhance the text. Chapter bibliographies. Appendices. Index. 468pp. ...x 8½. 0-486-64871-0

...ETHODS OF THERMODYNAMICS, Howard Reiss. Outstanding text focuses on ...hysical technique of thermodynamics, typical problem areas of understanding, and significance and use of thermodynamic potential. 1965 edition. 238pp. 5⅜ x 8½.
0-486-69445-3

THE ELECTROMAGNETIC FIELD, Albert Shadowitz. Comprehensive under- graduate text covers basics of electric and magnetic fields, builds up to electromagnetic theory. Also related topics, including relativity. Over 900 problems. 768pp. 5⅝ x 8¼.
0-486-65660-8

GREAT EXPERIMENTS IN PHYSICS: FIRSTHAND ACCOUNTS FROM GALILEO TO EINSTEIN, Morris H. Shamos (ed.). 25 crucial discoveries: Newton's laws of motion, Chadwick's study of the neutron, Hertz on electromagnetic waves, more. Original accounts clearly annotated. 370pp. 5⅝ x 8½. 0-486-25346-5

EINSTEIN'S LEGACY, Julian Schwinger. A Nobel Laureate relates fascinating story of Einstein and development of relativity theory in well-illustrated, nontechnical volume. Subjects include meaning of time, paradoxes of space travel, gravity and its effect on light, non-Euclidean geometry and curving of space-time, impact of radio astronomy and space-age discoveries, and more. 189 b/w illustrations. xiv+250pp. 8⅜ x 9¼. 0-486-41974-6

THE VARIATIONAL PRINCIPLES OF MECHANICS, Cornelius Lanczos. Philosophic, less formalistic approach to analytical mechanics offers model of clear, scholarly exposition at graduate level with coverage of basics, calculus of variations, principle of virtual work, equations of motion, more. 418pp. 5⅜ x 8½. 0-486-65067-7

Paperbound unless otherwise indicated. Available at your book dealer, online at www.doverpublications.com, or by writing to Dept. GI, Dover Publications, Inc., 31 East 2nd Street, Mineola, NY 11501. For current price information or for free catalogues (please indicate field of interest), write to Dover Publications or log on to www.doverpublications.com and see every Dover book in print. Dover publishes more than 400 books each year on science, elementary and advanced mathematics, biology, music, art, literary history, social sciences, and other areas.